ESSAYS

ON THE

THEORY OF NUMBERS

I. CONTINUITY AND IRRATIONAL NUMBERS
II. THE NATURE AND MEANING OF NUMBERS

BY

RICHARD DEDEKIND

AUTHORIZED TRANSLATION BY

WOOSTER WOODRUFF BEMAN
LATE PROFESSOR OF MATHEMATICS
THE UNIVERSITY OF MICHIGAN

DOVER PUBLICATIONS, INC.
NEW YORK

This new Dover edition, first published in 1963, is an unabridged and unaltered republication of the English translation first published by The Open Court Publishing Company in 1901.

International Standard Book Number: 0-486-21010-3
Library of Congress Catalog Card Number: 63-3681

Manufactured in the United States of America

Dover Publications, Inc.
180 Varick Street
New York 14, N. Y.

CONTENTS

I. CONTINUITY AND IRRATIONAL NUMBERS.

II. THE NATURE AND MEANING OF NUMBERS.

CONTINUITY AND IRRATIONAL NUMBERS

CONTINUITY AND IRRATIONAL NUMBERS.

MY attention was first directed toward the considerations which form the subject of this pamphlet in the autumn of 1858. As professor in the Polytechnic School in Zürich I found myself for the first time obliged to lecture upon the elements of the differential calculus and felt more keenly than ever before the lack of a really scientific foundation for arithmetic. In discussing the notion of the approach of a variable magnitude to a fixed limiting value, and especially in proving the theorem that every magnitude which grows continually, but not beyond all limits, must certainly approach a limiting value, I had recourse to geometric evidences. Even now such resort to geometric intuition in a first presentation of the differential calculus, I regard as exceedingly useful, from the didactic standpoint, and indeed indispensable, if one does not wish to lose too much time. But that this form of introduction into the differential calculus can make no claim to being scientific, no one will deny. For myself this feeling of dissatisfaction was so overpowering that I made the fixed resolve to keep meditating on the question till I should find a purely arithmetic and perfectly rigorous foundation for the principles of infinitesimal analysis. The statement is so frequently made that the differential calculus deals with continuous magnitude, and yet an explanation of this continuity is nowhere given; even the most rigorous expositions of the differential calculus do not base their proofs upon continuity but, with more or less consciousness of the fact, they either appeal to geometric notions or those suggested by geometry, or depend upon theorems which are never established in a purely arithmetic manner. Among these, for example, belongs the above-mentioned theorem, and a more careful investigation convinced me that this theorem, or any one equivalent to it, can be regarded in some way as a sufficient basis for infinitesimal analysis. It then only remained to discover its true origin in the elements of arithmetic and thus at the same time to secure a real definition of the essence of continuity. I succeeded Nov. 24, 1858, and a few days afterward I communicated the results of my meditations to my dear friend Durège with whom I had a long and lively discussion. Later I explained these views of a scientific basis of arithmetic to a few of my pupils, and here in Braunschweig read a paper upon the subject before the scientific club of professors, but I could not make up my mind to its publication, because in the first place, the presentation did not seem altogether simple, and further, the theory itself had little promise. Nevertheless I had already half determined to select this theme as subject for this occasion, when a few days ago, March 14, by the kindness of the author, the paper *Die Elemente der Funktionenlehre* by E. Heine (*Crelle's Journal,* Vol. 74) came into my hands and confirmed me in my decision. In the main I fully agree with the substance of this memoir, and indeed I could hardly do otherwise, but I will frankly acknowledge that my own presentation seems to me to be simpler in form and to bring out the vital point more clearly. While writing this preface (March 20, 1872), I am just in receipt of the interesting paper *Ueber die Ausdehnung eines Satzes aus der Theorie der trigonometrischen Reihen,* by G. Cantor (*Math. Annalen,* Vol. 5), for which I owe the ingenious author my hearty thanks. As I find on a hasty perusal, the axiom given in Section II. of that paper, aside from the form of presentation, agrees with what I designate in Section III. as the essence of continuity. But what advantage will be gained by even a purely abstract definition of real numbers of a higher type, I am as yet unable to see, conceiving as I do of the domain of real numbers as complete in itself.

I.

PROPERTIES OF RATIONAL NUMBERS.

The development of the arithmetic of rational numbers is here presupposed, but still I think it worth while to call attention to certain important matters without discussion, so as to show at the outset the standpoint assumed in what follows. I regard the whole of arithmetic as a necessary, or at least natural, consequence of the simplest arithmetic act, that of counting, and counting itself as nothing else than the successive creation of the infinite series of positive integers in which each individual is defined by the one immediately preceding; the simplest act is the passing from an already-formed individual to the consecutive new one to be formed. The chain of these numbers forms in itself an exceedingly useful instrument for the human mind; it presents an inexhaustible wealth of remarkable laws obtained by the introduction of the four fundamental operations of arithmetic. Addition is the combination of any arbitrary repetitions of the above-mentioned simplest act into a single act; from it in a similar way arises multiplication. While the performance of these two operations is always possible, that of the inverse operations, subtraction and division, proves to be limited. Whatever the immediate occasion may have been, whatever comparisons or analogies with experience, or intuition, may have led thereto; it is certainly true that just this limitation in performing the indirect operations has in each case been the real motive for a new creative act; thus negative and fractional numbers have been created by the human mind; and in the system of all rational numbers there has been gained an instrument of infinitely greater perfection. This system, which I shall denote by R, possesses first of all a completeness and self-containedness which I have designated in another place* as characteristic of a *body of numbers* [Zahlkörper] and which consists in this that the four fundamental operations are always performable with any two individuals in R, i. e., the result is always an individual of R, the single case of division by the number zero being excepted.

For our immediate purpose, however, another property of the system R is still more important; it may be expressed by saying that the system R forms a well-arranged domain of one dimension extending to infinity on two opposite sides. What is meant by this is sufficiently indicated by my use of expressions borrowed from geometric ideas; but just for this reason it will be necessary to bring out clearly the corresponding purely arithmetic properties in order to avoid even the appearance as if arithmetic were in need of ideas foreign to it.

To express that the symbols a and b represent one and the same rational number we put $a=b$ as well as $b=a$. The fact that two rational numbers a, b are different appears in this that the difference $a-b$ has either a positive or negative value. In the former case a is said to be *greater* than b, b *less* than a; this is also indicated by the symbols $a>b$, $b<a$.† As in the latter case $b-a$ has a positive value it follows

* *Vorlesungen über Zahlentheorie*, by P. G. Lejeune Dirichlet. 2d ed. § 159.

† Hence in what follows the so-called "algebraic" greater and less are understood unless the word "absolute" is added.

that $b > a$, $a < b$. In regard to these two ways in which two numbers may differ the following laws will hold:

I. If $a > b$, and $b > c$, then $a > c$. Whenever a, c are two different (or unequal) numbers, and b is greater than the one and less than the other, we shall, without hesitation because of the suggestion of geometric ideas, express this briefly by saying: b lies between the two numbers a, c.

II. If a, c are two different numbers, there are infinitely many different numbers lying between a, c.

III. If a is any definite number, then all numbers of the system R fall into two classes, A_1 and A_2, each of which contains infinitely many individuals; the first class A_1 comprises all numbers a_1 that are $< a$, the second class A_2 comprises all numbers a_2 that are $> a$; the number a itself may be assigned at pleasure to the first or second class, being respectively the greatest number of the first class or the least of the second. In every case the separation of the system R into the two classes A_1, A_2 is such that every number of the first class A_1 is less than every number of the second class A_2.

II.

COMPARISON OF THE RATIONAL NUMBERS WITH THE POINTS OF A STRAIGHT LINE.

The above-mentioned properties of rational numbers recall the corresponding relations of position of the points of a straight line L. If the two opposite directions existing upon it are distinguished by "right" and "left," and p, q are two different points, then either p lies to the right of q, and at the same time q to the left of p, or conversely q lies to the right of p and at the same time p to the left of q. A third case is impossible, if p, q are actually different points. In regard to this difference in position the following laws hold:

I. If p lies to the right of q, and q to the right of r, then p lies to the right of r; and we say that q lies between the points p and r.

II. If p, r are two different points, then there always exist infinitely many points that lie between p and r.

III. If p is a definite point in L, then all points in L fall into two classes, P_1, P_2, each of which contains infinitely many individuals; the first class P_1 contains all the points p_1, that lie to the left of p, and the second class P_2 contains all the points p_2 that lie to the right of p; the point p itself may be assigned at pleasure to the first or second class. In every case the separation of the straight line L into the two classes or portions P_1, P_2, is of such a character that every point of the first class P_1 lies to the left of every point of the second class P_2.

This analogy between rational numbers and the points of a straight line, as is well known, becomes a real correspondence when we select upon the straight line a definite origin or zero-point o and a definite unit of length for the measurement of segments. With the aid of the latter to every rational number a a corresponding length can be constructed and if we lay this off upon the straight line to the right or left of o according as a is positive or negative, we obtain a definite end-point p, which may be regarded as the point corresponding to the number a; to the rational number zero corresponds the point o. In this way to every rational number a, i. e., to every individual in R, corresponds one and only one point p, i. e., an individual in L. To the two numbers a, b respectively correspond the two points p, q, and if $a > b$, then p lies to the right of q. To the laws I, II, III of the previous Section correspond completely the laws I, II, III of the present.

III.

CONTINUITY OF THE STRAIGHT LINE.

Of the greatest importance, however, is the fact that in the straight line L there are infinitely many points which correspond to no rational number. If the point p corresponds to the rational number a, then, as is well known, the length op is commensurable with the invariable unit of measure used in the construction, i. e., there exists a third length, a so-called common measure, of which these two lengths are integral multiples. But the ancient Greeks already knew and had demonstrated that there are lengths incommensurable with a given unit of length, e. g., the diagonal of the square whose side is the unit of length. If we lay off such a length from the point o upon the line we obtain an end-point which corresponds to no rational number. Since further it can be easily shown that there are infinitely many lengths which are incommensurable with the unit of length, we may affirm: The straight line L is infinitely richer in point-individuals than the domain R of rational numbers in number-individuals.

If now, as is our desire, we try to follow up arithmetically all phenomena in the straight line, the domain of rational numbers is insufficient and it becomes absolutely necessary that the instrument R constructed by the creation of the rational numbers be essentially improved by the creation of new numbers such that the domain of numbers shall gain the same completeness, or as we may say at once, the same *continuity*, as the straight line.

The previous considerations are so familiar and well known to all that many will regard their repetition quite superfluous. Still I regarded this recapitulation as necessary to prepare properly for the main question. For, the way in which the irrational numbers are usually introduced is based directly upon the conception of extensive magnitudes—which itself is nowhere carefully defined—and explains number as the result of measuring such a magnitude by another of the same kind.* Instead of this I demand that arithmetic shall be developed out of itself.

That such comparisons with non-arithmetic notions have furnished the immediate occasion for the extension of the number-concept may, in a general way, be granted (though this was certainly not the case in the introduction of complex numbers); but this surely is no sufficient ground for introducing these foreign notions into arithmetic, the science of numbers. Just as negative and fractional rational numbers are formed by a new creation, and as the laws of operating with these numbers must and can be reduced to the laws of operating with positive integers, so we must endeavor completely to define irrational numbers by means of the rational numbers alone. The question only remains how to do this.

The above comparison of the domain R of rational numbers with a straight line has led to the recognition of the existence of gaps, of a certain incompleteness or discontinuity of the former, while we ascribe to the straight line completeness, absence of gaps, or continuity. In what then does this continuity consist? Everything must depend on the answer to this question, and only through it shall we obtain a scientific basis for the investigation of *all* continuous domains. By vague remarks upon the unbroken connection in

*The apparent advantage of the generality of this definition of number disappears as soon as we consider complex numbers. According to my view, on the other hand, the notion of the ratio between two numbers of the same kind can be clearly developed only after the introduction of irrational numbers.

the smallest parts obviously nothing is gained; the problem is to indicate a precise characteristic of continuity that can serve as the basis for valid deductions. For a long time I pondered over this in vain, but finally I found what I was seeking. This discovery will, perhaps, be differently estimated by different people; the majority may find its substance very commonplace. It consists of the following. In the preceding section attention was called to the fact that every point p of the straight line produces a separation of the same into two portions such that every point of one portion lies to the left of every point of the other. I find the essence of continuity in the converse, i. e., in the following principle:

"If all points of the straight line fall into two classes such that every point of the first class lies to the left of every point of the second class, then there exists one and only one point which produces this division of all points into two classes, this severing of the straight line into two portions."

As already said I think I shall not err in assuming that every one will at once grant the truth of this statement; the majority of my readers will be very much disappointed in learning that by this commonplace remark the secret of continuity is to be revealed. To this I may say that I am glad if every one finds the above principle so obvious and so in harmony with his own ideas of a line; for I am utterly unable to adduce any proof of its correctness, nor has any one the power. The assumption of this property of the line is nothing else than an axiom by which we attribute to the line its continuity, by which we find continuity in the line. If space has at all a real existence it is *not* necessary for it to be continuous; many of its properties would remain the same even were it discontinuous. And if we knew for certain that space was discontinuous there would be nothing to prevent us, in case we so desired, from filling up its gaps, in thought, and thus making it continuous; this filling up would consist in a creation of new point-individuals and would have to be effected in accordance with the above principle.

IV.

CREATION OF IRRATIONAL NUMBERS.

From the last remarks it is sufficiently obvious how the discontinuous domain R of rational numbers may be rendered complete so as to form a continuous domain. In Section I it was pointed out that every rational number a effects a separation of the system R into two classes such that every number a_1 of the first class A_1 is less than every number a_2 of the second class A_2; the number a is either the greatest number of the class A_1 or the least number of the class A_2. If now any separation of the system R into two classes A_1, A_2, is given which possesses only *this* characteristic property that every number a_1 in A_1 is less than every number a_2 in A_2, then for brevity we shall call such a separation a *cut* [Schnitt] and designate it by (A_1, A_2). We can then say that every rational number a produces one cut or, strictly speaking, two cuts, which, however, we shall not look upon as essentially different; this cut possesses, *besides*, the property that either among the numbers of the first class there exists a greatest or among the numbers of the second class a least number. And conversely, if a cut possesses this property, then it is produced by this greatest or least rational number.

But it is easy to show that there exist infinitely many cuts not produced by rational numbers. The following example suggests itself most readily.

Let D be a positive integer but not the square of an integer, then there exists a positive integer λ such that

$$\lambda^2 < D < (\lambda + 1)^2.$$

If we assign to the second class A_2, every positive rational number a_2 whose square is $> D$, to the first class A_1 all other rational numbers a_1, this separation forms a cut (A_1, A_2), i. e., every number a_1 is less than every number a_2. For if $a_1 = 0$, or is negative, then on that ground a_1 is less than any number a_2, because, by definition, this last is positive; if a_1 is positive, then is its square $\leqq D$, and hence a_1 is less than any positive number a_2 whose square is $> D$.

But this cut is produced by no rational number. To demonstrate this it must be shown first of all that there exists no rational number whose square $= D$.

Although this is known from the first elements of the theory of numbers, still the following indirect proof may find place here. If there exist a rational number whose square $= D$, then there exist two positive integers t, u, that satisfy the equation

$$t^2 - Du^2 = 0,$$

and we may assume that u is the *least* positive integer possessing the property that its square, by multiplication by D, may be converted into the square of an integer t. Since evidently

$$\lambda u < t < (\lambda + 1)u,$$

the number $u' = t - \lambda u$ is a positive integer certainly *less* than u. If further we put

$$t' = Du - \lambda t,$$

t' is likewise a positive integer, and we have

$$t'^2 - Du'^2 = (\lambda^2 - D)(t^2 - Du^2) = 0,$$

which is contrary to the assumption respecting u.

Hence the square of every rational number x is either $< D$ or $> D$. From this it easily follows that there is neither in the class A_1 a greatest, nor in the class A_2 a least number. For if we put

$$y = \frac{x(x^2 + 3D)}{3x^2 + D},$$

we have

$$y - x = \frac{2x(D - x^2)}{3x^2 + D}$$

and

$$y^2 - D = \frac{(x^2 - D)^3}{(3x^2 + D)^2}.$$

If in this we assume x to be a positive number from the class A_1, then $x^2 < D$, and hence $y > x$ and $y^2 < D$. Therefore y likewise belongs to the class A_1. But if we assume x to be a number from the class A_2, then $x^2 > D$, and hence $y < x$, $y > 0$, and $y^2 > D$. Therefore y likewise belongs to the class A_2. This cut is therefore produced by no rational number.

In this property that not all cuts are produced by rational numbers consists the incompleteness or discontinuity of the domain R of all rational numbers

Whenever, then, we have to do with a cut (A_1, A_2) produced by no rational number, we create a new, an *irrational* number α, which we regard as completely defined by this cut (A_1, A_2); we shall say that the number α corresponds to this cut, or that it produces this cut. From now on, therefore, to every definite cut there corresponds a definite rational or irrational number, and we regard two numbers as *different* or *unequal* always and only when they correspond to essentially different cuts.

In order to obtain a basis for the orderly arrangement of all *real*, i. e., of all rational and irrational numbers we must investigate the relation between any two cuts (A_1, A_2) and (B_1, B_2) produced by any two numbers α and β. Obviously a cut (A_1, A_2) is given completely when one of the two classes, e. g., the first A_1 is known, because the second A_2 consists of all rational numbers not contained in A_1, and the characteristic property of such a first class lies in this that if the number a_1 is contained in it, it also contains all numbers less than a_1. If now we compare two such first classes A_1, B_1 with each other, it may happen

1. That they are perfectly identical, i. e., that every number contained in A_1 is also contained in B_1, and that every number contained in B_1 is also contained in A_1. In this case A_2 is necessarily identical with B_2, and the two cuts are perfectly identical, which we denote in symbols by $\alpha = \beta$ or $\beta = \alpha$.

But if the two classes A_1, B_1 are not identical, then there exists in the one, e. g., in A_1, a number $a'_1 = b'_2$ not contained in the other B_1 and consequently found in B_2; hence all numbers b_1 contained in B_1 are certainly less than this number $a'_1 = b'_2$ and therefore all numbers b_1 are contained in A_1.

2. If now this number a'_1 is the only one in A_1 that is not contained in B_1, then is every other number a_1 contained in A_1 also contained in B_1 and is consequently $< a'_1$, i. e., a'_1 is the greatest among all the numbers a_1, hence the cut (A_1, A_2) is produced by the rational number $\alpha = a'_1 = b'_2$. Concerning the other cut (B_1, B_2) we know already that all numbers b_1 in B_1 are also contained in A_1 and are less than the number $a'_1 = b'_2$ which is contained in B_2; every other number b_2 contained in B_2 must, however, be greater than b'_2, for otherwise it would be less than a'_1, therefore contained in A_1 and hence in B_1; hence b'_2 is the least among all numbers contained in B_2,

and consequently the cut (B_1, B_2) is produced by the same rational number $\beta = b'_2 = a'_1 = \alpha$. The two cuts are then only unessentially different.

3. If, however, there exist in A_1 at least two different numbers $a'_1 = b'_2$ and $a''_1 = b''_2$, which are not contained in B_1, then there exist infinitely many of them, because all the infinitely many numbers lying between a'_1 and a''_1 are obviously contained in A_1 (Section I, II) but not in B_1. In this case we say that the numbers α and β corresponding to these two essentially different cuts (A_1, A_2) and (B_1, B_2) are *different*, and further that α is *greater* than β, that β is *less* than α, which we express in symbols by $\alpha > \beta$ as well as $\beta < \alpha$. It is to be noticed that this definition coincides completely with the one given earlier, when α, β are rational.

The remaining possible cases are these:

4. If there exists in B_1 one and only one number $b'_1 = a'_2$, that is not contained in A_1 then the two cuts (A_1, A_2) and (B_1, B_2) are only unessentially different and they are produced by one and the same rational number $\alpha = a'_2 = b'_1 = \beta$.

5. But if there are in B_1 at least two numbers which are not contained in A_1, then $\beta > \alpha$, $\alpha < \beta$.

As this exhausts the possible cases, it follows that of two different numbers one is necessarily the greater, the other the less, which gives two possibilities. A third case is impossible. This was indeed involved in the use of the *comparative* (greater, less) to designate the relation between α, β; but this use has only now been justified. In just such investigations one needs to exercise the greatest care so that even with the best intention to be honest he shall not, through a hasty choice of expressions borrowed from other notions already developed, allow himself to be led into the use of inadmissible transfers from one domain to the other.

If now we consider again somewhat carefully the case $\alpha > \beta$ it is obvious that the less number β, if rational, certainly belongs to the class A_1; for since there is in A_1 a number $a'_1 = b'_2$ which belongs to the class B_2, it follows that the number β, whether the greatest number in B_1 or the least in B_2 is certainly $\leqq a'_1$ and hence contained in A_1. Likewise it is obvious from $\alpha > \beta$ that the greater number α, if rational, certainly belongs to the class B_2, because $\alpha \geqq a'_1$. Combining these two considerations we get the following result: If a cut is produced by the number α then any rational number belongs to the class A_1 or to the class A_2 according as it is less or greater than α; if the number α is itself rational it may belong to either class.

From this we obtain finally the following: If $\alpha > \beta$, i. e., if there are infinitely many numbers in A_1 not contained in B_1 then there are infinitely many such numbers that at the same time are different from α and from β; every such rational number c is $< \alpha$, because it is contained in A_1 and at the same time it is $> \beta$ because contained in B_2.

V.

CONTINUITY OF THE DOMAIN OF REAL NUMBERS.

In consequence of the distinctions just established the system $\mathfrak{R}$ of all real numbers forms a well-arranged domain of one dimension; this is to mean merely that the following laws prevail:

I. If $\alpha > \beta$, and $\beta > \gamma$, then is also $\alpha > \gamma$. We shall say that the number β lies between α and γ.

II. If α, γ are any two different numbers, then there exist infinitely many different numbers β lying between α, γ.

III. If α is any definite number then all numbers of the system $\mathfrak{R}$ fall into two classes $\mathfrak{A}_1$ and $\mathfrak{A}_2$ each of which contains infinitely many individuals; the first class $\mathfrak{A}_1$ comprises all the numbers α_1 that are less than α, the second $\mathfrak{A}_2$ comprises all the numbers α_2 that are greater than α; the number α itself may be assigned at pleasure to the first class or to the second, and it is respectively the greatest of the first or the least of the second class. In each case the separation of the system $\mathfrak{R}$ into the two classes $\mathfrak{A}_1$, $\mathfrak{A}_2$ is such that every number of the first class $\mathfrak{A}_1$ is smaller than every number of the second class $\mathfrak{A}_2$ and we say that this separation is produced by the number α.

For brevity and in order not to weary the reader I suppress the proofs of these theorems which follow immediately from the definitions of the previous section.

Beside these properties, however, the domain $\mathfrak{R}$ possesses also *continuity*; i. e., the following theorem is true:

IV. If the system $\mathfrak{R}$ of all real numbers breaks up into two classes $\mathfrak{A}_1$, $\mathfrak{A}_2$ such that every number α_1 of the class $\mathfrak{A}_1$ is less than every number α_2 of the class $\mathfrak{A}_2$ then there exists one and only one number α by which this separation is produced.

Proof. By the separation or the cut of $\mathfrak{R}$ into $\mathfrak{A}_1$ and $\mathfrak{A}_2$ we obtain at the same time a cut (A_1, A_2) of the system R of all rational numbers which is defined by this that A_1 contains all rational numbers of the class $\mathfrak{A}_1$ and A_2 all other rational numbers, i. e., all rational numbers of the class $\mathfrak{A}_2$. Let α be the perfectly definite number which produces this cut (A_1, A_2). If β is any number different from α, there are always infinitely many rational numbers c lying between α and β. If $\beta < \alpha$, then $c < \alpha$; hence c belongs to the class A_1 and consequently also to the class $\mathfrak{A}_1$, and since at the same time $\beta < c$ then β also belongs to the same class $\mathfrak{A}_1$, because every number in $\mathfrak{A}_2$ is greater than every number c in $\mathfrak{A}_1$. But if $\beta > \alpha$, then is $c > \alpha$; hence c belongs to the class A_2 and consequently also to the class $\mathfrak{A}_2$, and since at the same time $\beta > c$, then β also belongs to the same class $\mathfrak{A}_2$, because every number in $\mathfrak{A}_1$ is less than every number c in $\mathfrak{A}_2$. Hence every number β different from α belongs to the class $\mathfrak{A}_1$ or to the class $\mathfrak{A}_2$ according as $\beta < \alpha$ or $\beta > \alpha$; consequently α itself is either the greatest number in $\mathfrak{A}_1$ or the least number in $\mathfrak{A}_2$, i. e., α is one and obviously the only number by which the separation of R into the classes $\mathfrak{A}_1$, $\mathfrak{A}_2$ is produced. Which was to be proved.

VI.

OPERATIONS WITH REAL NUMBERS.

To reduce any operation with two real numbers α, β to operations with rational numbers, it is only necessary from the cuts (A_1, A_2), (B_1, B_2) produced by the numbers α and β in the system R to define the cut (C_1, C_2) which is to correspond to the result of the operation, γ. I confine myself here to the discussion of the simplest case, that of addition.

If c is any rational number, we put it into the class C_1, provided there are two numbers one a_1 in A_1 and one b_1 in B_1 such that their sum $a_1 + b_1 \geqq c$; all other rational numbers shall be put into the class C_2. This separation of all rational numbers into the two classes C_1, C_2 evidently forms a cut, since every number c_1 in C_1 is less than every number c_2 in C_2. If both α and β are rational, then every number c_1 contained in C_1 is $\leqq \alpha + \beta$, because $a_1 \leqq \alpha$, $b_1 \leqq \beta$, and therefore $a_1 + b_1 \leqq \alpha + \beta$; further, if there were contained in C_2 a number $c_2 < \alpha + \beta$, hence $\alpha + \beta = c_2 + p$, where p is a positive rational number, then we should have

$$c_2 = (\alpha - \tfrac{1}{2}p) + (\beta - \tfrac{1}{2}p),$$

which contradicts the definition of the number c_2, because $\alpha - \frac{1}{2}p$ is a number in A_1, and $\beta - \frac{1}{2}p$ a number in B_1; consequently every number c_2 contained in C_2 is $\geqq \alpha + \beta$. Therefore in this case the cut (C_1, C_2) is produced by the sum $\alpha + \beta$. Thus we shall not violate the definition which holds in the arithmetic of rational numbers if in all cases we understand by the sum $\alpha + \beta$ of any two real numbers α, β that number γ by which the cut (C_1, C_2) is produced. Further, if only one of the two numbers α, β is rational, e. g., α, it is easy to see that it makes no difference with the sum $\gamma = \alpha + \beta$ whether the number α is put into the class A_1 or into the class A_2.

Just as addition is defined, so can the other operations of the so-called elementary arithmetic be defined, viz., the formation of differences, products, quotients, powers, roots, logarithms, and in this way we arrive at real proofs of theorems (as, e. g., $\sqrt{2} \cdot \sqrt{3} = \sqrt{6}$), which to the best of my knowledge have never been established before. The excessive length that is to be feared in the definitions of the more complicated operations is partly inherent in the nature of the subject but can for the most part be avoided. Very useful in this connection is the notion of an *interval*, i. e., a system A of rational numbers possessing the following characteristic property: if a and a' are numbers of the system A, then are all rational numbers lying between a and a' contained in A. The system R of all rational numbers, and also the two classes of any cut are intervals. If there exist a rational number a_1 which is less and a rational number a_2 which is greater than every number of the interval A, then A is called a finite interval; there then exist infinitely many numbers in the same condition as a_1 and infinitely many in the same condition as a_2; the whole domain R breaks up into three parts A_1, A, A_2 and there enter two perfectly definite rational or irrational numbers α_1, α_2 which may be called respectively the lower and upper (or the less and greater) *limits* of the interval; the lower limit α_1 is determined by the cut for which the system A_1 forms the first class and the upper α_2 by the cut for which the system A_2 forms the second class. Of every rational or irrational number α lying between α_1 and α_2 it may be said that it lies *within* the interval A. If all numbers of an interval A are also numbers of an interval B, then A is called a portion of B.

Still lengthier considerations seem to loom up when we attempt to adapt the numerous theorems of the arithmetic of rational numbers (as, e. g., the theorem $(a+b)c=ac+bc$) to any real numbers. This, however, is not the case. It is easy to see that it all reduces to showing that the arithmetic operations possess a certain continuity. What I mean by this statement may be expressed in the form of a general theorem:

"If the number λ is the result of an operation performed on the numbers α, β, γ, . . . and λ lies within the interval L, then intervals A, B, C, . . . can be taken within which lie the numbers α, β, γ, . . . such that the result of the same operation in which the numbers α, β, γ, . . . are replaced by arbitrary numbers of the intervals A, B, C, . . . is always a number lying within the interval L." The forbidding clumsiness, however, which marks the statement of such a theorem convinces us that something must be brought in as an aid to expression; this is, in fact, attained in the most satisfactory way by introducing the ideas of *variable magnitudes*, *functions*, *limiting values*, and it would be best to base the definitions of even the simplest arithmetic operations upon these ideas, a matter which, however, cannot be carried further here.

VII.

INFINITESIMAL ANALYSIS.

Here at the close we ought to explain the connection between the preceding investigations and certain fundamental theorems of infinitesimal analysis.

We say that a variable magnitude x which passes through successive definite numerical values approaches a fixed limiting value α when in the course of the process x lies finally between two numbers between which α itself lies, or, what amounts to the same, when the difference $x - \alpha$ taken absolutely becomes finally less than any given value different from zero.

One of the most important theorems may be stated in the following manner: "If a magnitude x grows continually but not beyond all limits it approaches a limiting value."

I prove it in the following way. By hypothesis there exists one and hence there exist infinitely many numbers α_2 such that x remains continually $< \alpha_2$; I designate by $\mathfrak{A}_2$ the system of all these numbers α_2, by $\mathfrak{A}_1$ the system of all other numbers α_1; each of the latter possesses the property that in the course of the process x becomes finally $\geqq \alpha_1$, hence every number α_1 is less than every number α_2 and consequently there exists a number α which is either the greatest in $\mathfrak{A}_1$ or the least in $\mathfrak{A}_2$ (V, IV). The former cannot be the case since x never ceases to grow, hence α is the least number in $\mathfrak{A}_2$ Whatever number α_1 be taken we shall have finally $\alpha_1 < x < \alpha$, i. e., x approaches the limiting value α.

This theorem is equivalent to the principle of continuity, i. e., it loses its validity as soon as we assume a single real number not to be contained in the domain $\mathfrak{R}$; or otherwise expressed: if this theorem is correct, then is also theorem IV. in V. correct.

Another theorem of infinitesimal analysis, likewise equivalent to this, which is still oftener employed, may be stated as follows: "If in the variation of a magnitude x we can for every given positive magnitude δ assign a corresponding position from and after which x changes by less than δ then x approaches a limiting value."

This converse of the easily demonstrated theorem

that every variable magnitude which approaches a limiting value finally changes by less than any given positive magnitude can be derived as well from the preceding theorem as directly from the principle of continuity. I take the latter course. Let δ be any positive magnitude (i. e., $\delta > 0$), then by hypothesis a time will come after which x will change by less than δ, i. e., if at this time x has the value a, then afterwards we shall continually have $x > a - \delta$ and $x < a + \delta$. I now for a moment lay aside the original hypothesis and make use only of the theorem just demonstrated that all later values of the variable x lie between two assignable finite values. Upon this I base a double separation of all real numbers. To the system $\mathfrak{A}_2$ I assign a number α_2 (e. g., $a + \delta$) when in the course of the process x becomes finally $\leqq \alpha_2$; to the system $\mathfrak{A}_1$ I assign every number not contained in $\mathfrak{A}_2$; if α_1 is such a number, then, however far the process may have advanced, it will still happen infinitely many times that $x > \alpha_2$. Since every number α_1 is less than every number α_2 there exists a perfectly definite number α which produces this cut $(\mathfrak{A}_1, \mathfrak{A}_2)$ of the system $\mathfrak{R}$ and which I will call the upper limit of the variable x which always remains finite. Likewise as a result of the behavior of the variable x a second cut $(\mathfrak{B}_1, \mathfrak{B}_2)$ of the system $\mathfrak{R}$ is produced; a number β_2 (e.g., $a - \delta$) is assigned to $\mathfrak{B}_2$ when in the course of the process x becomes finally $\geqq \beta$; every other number β_2, to be assigned to $\mathfrak{B}_2$, has the property that x is never finally $\geqq \beta_2$; therefore infinitely many times x becomes $< \beta_2$; the number β by which this cut is produced I call the lower limiting value of the variable x. The two numbers α, β are obviously characterised by the following property: if ϵ is an arbitrarily small positive magnitude then we have always finally $x < \alpha + \epsilon$ and $x > \beta - \epsilon$, but never finally $x < \alpha - \epsilon$ and never finally $x > \beta + \epsilon$. Now two cases are possible. If α and β are different from each other, then necessarily $\alpha > \beta$, since continually $\alpha_2 \geqq \beta_2$; the variable x oscillates, and, however far the process advances, always undergoes changes whose amount surpasses the value $(\alpha - \beta) - 2\epsilon$ where ϵ is an arbitrarily small positive magnitude. The original hypothesis to which I now return contradicts this consequence; there remains only the second case $\alpha = \beta$ and since it has already been shown that, however small be the positive magnitude ϵ, we always have finally $x < \alpha + \epsilon$ and $x > \beta - \epsilon$, x approaches the limiting value α, which was to be proved.

These examples may suffice to bring out the connection between the principle of continuity and infinitesimal analysis.

THE NATURE AND MEANING OF NUMBERS

PREFACE TO THE FIRST EDITION.

IN science nothing capable of proof ought to be accepted without proof. Though this demand seems so reasonable yet I cannot regard it as having been met even in the most recent methods of laying the foundations of the simplest science; viz., that part of logic which deals with the theory of numbers.* In speaking of arithmetic (algebra, analysis) as a part of logic I mean to imply that I consider the number-concept entirely independent of the notions or intuitions of space and time, that I consider it an immediate result from the laws of thought. My answer to the problems propounded in the title of this paper is, then, briefly this: numbers are free creations of the human mind; they serve as a means of apprehending more easily and more sharply the difference of things. It is only through the purely logical process of building up the science of numbers and by thus acquiring

*Of the works which have come under my observation I mention the valuable *Lehrbuch der Arithmetik und Algebra* of E. Schröder (Leipzig, 1873), which contains a bibliography of the subject, and in addition the memoirs of Kronecker and von Helmholtz upon the Number-Concept and upon Counting and Measuring (in the collection of philosophical essays published in honor of E. Zeller, Leipzig, 1887). The appearance of these memoirs has induced me to publish my own views in many respects similar but in foundation essentially different, which I formulated many years ago in absolute independence of the works of others.

the continuous number-domain that we are prepared accurately to investigate our notions of space and time by bringing them into relation with this number-domain created in our mind.* If we scrutinise closely what is done in counting an aggregate or number of things, we are led to consider the ability of the mind to relate things to things, to let a thing correspond to a thing, or to represent a thing by a thing, an ability without which no thinking is possible. Upon this unique and therefore absolutely indispensable foundation, as I have already affirmed in an announcement of this paper,† must, in my judgment, the whole science of numbers be established. The design of such a presentation I had formed before the publication of my paper on *Continuity*, but only after its appearance and with many interruptions occasioned by increased official duties and other necessary labors, was I able in the years 1872 to 1878 to commit to paper a first rough draft which several mathematicians examined and partially discussed with me. It bears the same title and contains, though not arranged in the best order, all the essential fundamental ideas of my present paper, in which they are more carefully elaborated. As such main points I mention here the sharp distinction between finite and infinite (64), the notion of the number [*Anzahl*] of things (161), the

* See Section III. of my memoir, *Continuity and Irrational Numbers* (Braunschweig, 1872), translated at pages 8 et seq. of the present volume.

† Dirichlet's *Vorlesungen über Zahlentheorie*, third edition, 1879, § 163, note on page 470.

proof that the form of argument known as complete induction (or the inference from n to $n+1$) is really conclusive (59), (60), (80), and that therefore the definition by induction (or recursion) is determinate and consistent (126).

This memoir can be understood by any one possessing what is usually called good common sense; no technical philosophic, or mathematical, knowledge is in the least degree required. But I feel conscious that many a reader will scarcely recognise in the shadowy forms which I bring before him his numbers which all his life long have accompanied him as faithful and familiar friends; he will be frightened by the long series of simple inferences corresponding to our step-by-step understanding, by the matter-of-fact dissection of the chains of reasoning on which the laws of numbers depend, and will become impatient at being compelled to follow out proofs for truths which to his supposed inner consciousness seem at once evident and certain. On the contrary in just this possibility of reducing such truths to others more simple, no matter how long and apparently artificial the series of inferences, I recognise a convincing proof that their possession or belief in them is never given by inner consciousness but is always gained only by a more or less complete repetition of the individual inferences. I like to compare this action of thought, so difficult to trace on account of the rapidity of its performance, with the action which an accomplished reader performs in reading; this reading always remains a more or less complete repetition of the individual steps which the beginner has to take in his wearisome spelling-out; a very small part of the same, and therefore a very small effort or exertion of the mind, is sufficient for the practised reader to recognise the correct, true word, only with very great probability, to be sure; for, as is well known, it occasionally happens that even the most practised proof-reader allows a typographical error to escape him, i. e., reads falsely, a thing which would be impossible if the chain of thoughts associated with spelling were fully repeated. So from the time of birth, continually and in increasing measure we are led to relate things to things and thus to use that faculty of the mind on which the creation of numbers depends; by this practice continually occurring, though without definite purpose, in our earliest years and by the attending formation of judgments and chains of reasoning we acquire a store of real arithmetic truths to which our first teachers later refer as to something simple, self-evident, given in the inner consciousness; and so it happens that many very complicated notions (as for example that of the number [*Anzahl*] of things) are erroneously regarded as simple. In this sense which I wish to express by the word formed after a well-known saying ἀεὶ ὁ ἄνθρωπος ἀριθμητίζει, I hope that the following pages, as an attempt to establish the science of numbers upon a uniform foundation will find a generous welcome and that other mathematicians will be led to reduce the long series of inferences to more moderate and attractive proportions.

In accordance with the purpose of this memoir I restrict myself to the consideration of the series of so-called natural numbers. In what way the gradual extension of the number-concept, the creation of zero, negative, fractional, irrational and complex numbers are to be accomplished by reduction to the earlier notions and that without any introduction of foreign conceptions (such as that of measurable magnitudes, which according to my view can attain perfect clearness only through the science of numbers), this I have shown at least for irrational numbers in my former memoir on *Continuity* (1872); in a way wholly similar, as I have already shown in Section III. of that memoir,* may the other extensions be treated, and I propose sometime to present this whole subject in systematic form. From just this point of view it appears as something self-evident and not new that every theorem of algebra and higher analysis, no matter how remote, can be expressed as a theorem about natural numbers,—a declaration I have heard repeatedly from the lips of Dirichlet. But I see nothing meritorious—and this was just as far from Dirichlet's thought—in actually performing this wearisome circumlocution and insisting on the use and recognition of no other than rational numbers. On the contrary,

*Pages 8 et seq. of the present volume.

the greatest and most fruitful advances in mathematics and other sciences have invariably been made by the creation and introduction of new concepts, rendered necessary by the frequent recurrence of complex phenomena which could be controlled by the old notions only with difficulty. On this subject I gave a lecture before the philosophic faculty in the summer of 1854 on the occasion of my admission as privat-docent in Göttingen. The scope of this lecture met with the approval of Gauss; but this is not the place to go into further detail.

Instead of this I will use the opportunity to make some remarks relating to my earlier work, mentioned above, on *Continuity and Irrational Numbers*. The theory of irrational numbers there presented, wrought out in the fall of 1853, is based on the phenomenon (Section IV.)* occurring in the domain of rational numbers which I designate by the term cut [*Schnitt*] and which I was the first to investigate carefully; it culminates in the proof of the continuity of the new domain of real numbers (Section V., iv.).† It appears to me to be somewhat simpler, I might say easier, than the two theories, different from it and from each other, which have been proposed by Weierstrass and G. Cantor, and which likewise are perfectly rigorous. It has since been adopted without essential modification by U. Dini in his *Fondamenti per la teorica delle*

*Pages 12 et seq. of the present volume.

†Page 20 of the present volume.

funzioni di variabili reali (Pisa, 1878); but the fact that in the course of this exposition my name happens to be mentioned, not in the description of the purely arithmetic phenomenon of the cut but when the author discusses the existence of a measurable quantity corresponding to the cut, might easily lead to the supposition that my theory rests upon the consideration of such quantities. Nothing could be further from the truth; rather have I in Section III.* of my paper advanced several reasons why I wholly reject the introduction of measurable quantities; indeed, at the end of the paper I have pointed out with respect to their existence that for a great part of the science of space the continuity of its configurations is not even a necessary condition, quite aside from the fact that in works on geometry arithmetic is only casually mentioned by name but is never clearly defined and therefore cannot be employed in demonstrations. To explain this matter more clearly I note the following example: If we select three non-collinear points A, B, C at pleasure, with the single limitation that the ratios of the distances AB, AC, BC are algebraic numbers,† and regard as existing in space only those points M, for which the ratios of AM, BM, CM to AB are likewise algebraic numbers, then is the space made up of the points M, as is easy to see, everywhere dis-

*Pages 8 et seq. of the present volume.

† Dirichlet's *Vorlesungen über Zahlentheorie*, § 159 of the second edition, § 160 of the third.

continuous; but in spite of this discontinuity, and despite the existence of gaps in this space, all constructions that occur in Euclid's *Elements*, can, so far as I can see, be just as accurately effected as in perfectly continuous space; the discontinuity of this space would not be noticed in Euclid's science, would not be felt at all. If any one should say that we cannot conceive of space as anything else than continuous, I should venture to doubt it and to call attention to the fact that a far advanced, refined scientific training is demanded in order to perceive clearly the essence of continuity and to comprehend that besides rational quantitative relations, also irrational, and besides algebraic, also transcendental quantitative relations are conceivable. All the more beautiful it appears to me that without any notion of measurable quantities and simply by a finite system of simple thought-steps man can advance to the creation of the pure continuous number-domain; and only by this means in my view is it possible for him to render the notion of continuous space clear and definite.

The same theory of irrational numbers founded upon the phenomenon of the cut is set forth in the *Introduction à la théorie des fonctions d'une variable* by J. Tannery (Paris, 1886). If I rightly understand a passage in the preface to this work, the author has thought out his theory independently, that is, at a time when not only my paper, but Dini's *Fondamenti* mentioned in the same preface, was unknown to him.

This agreement seems to me a gratifying proof that my conception conforms to the nature of the case, a fact recognised by other mathematicians, e. g., by Pasch in his *Einleitung in die Differential- und Integralrechnung* (Leipzig, 1883). But I cannot quite agree with Tannery when he calls this theory the development of an idea due to J. Bertrand and contained in his *Traité d'arithmétique*, consisting in this that an irrational number is defined by the specification of all rational numbers that are less and all those that are greater than the number to be defined. As regards this statement which is repeated by Stolz—apparently without careful investigation—in the preface to the second part of his *Vorlesungen über allgemeine Arithmetik* (Leipzig, 1886), I venture to remark the following: That an irrational number is to be considered as fully defined by the specification just described, this conviction certainly long before the time of Bertrand was the common property of all mathematicians who concerned themselves with the notion of the irrational. Just this manner of determining it is in the mind of every computer who calculates the irrational root of an equation by approximation, and if, as Bertrand does exclusively in his book, (the eighth edition, of the year 1885, lies before me,) one regards the irrational number as the ratio of two measurable quantities, then is this manner of determining it already set forth in the clearest possible way in the celebrated definition which Euclid gives of the equality of two ratios (*Elements*, V., 5). This same most ancient conviction has been the source of my theory as well as that of Bertrand and many other more or less complete attempts to lay the foundations for the introduction of irrational numbers into arithmetic. But though one is so far in perfect agreement with Tannery, yet in an actual examination he cannot fail to observe that Bertrand's presentation, in which the phenomenon of the cut in its logical purity is not even mentioned, has no similarity whatever to mine, inasmuch as it resorts at once to the existence of a measurable quantity, a notion which for reasons mentioned above I wholly reject. Aside from this fact this method of presentation seems also in the succeeding definitions and proofs, which are based on the postulate of this existence, to present gaps so essential that I still regard the statement made in my paper (Section VI.),* that the theorem $\sqrt{2} \cdot \sqrt{3} = \sqrt{6}$ has nowhere yet been strictly demonstrated, as justified with respect to this work also, so excellent in many other regards and with which I was unacquainted at that time.

R. DEDEKIND.

HARZBURG, October 5, 1887.

* Pages 21 et seq. of this volume.

PREFACE TO THE SECOND EDITION.

THE present memoir soon after its appearance met with both favorable and unfavorable criticisms; indeed serious faults were charged against it. I have been unable to convince myself of the justice of these charges, and I now issue a new edition of the memoir, which for some time has been out of print, without change, adding only the following notes to the first preface.

The property which I have employed as the defi nition of the infinite system had been pointed out before the appearance of my paper by G. Cantor (*Ein Beitrag zur Mannigfaltigkeitslehre*, *Crelle's Journal*, Vol. 84, 1878), as also by Bolzano (*Paradoxien des Unendlichen*, § 20, 1851). But neither of these authors made the attempt to use this property for the definition of the infinite and upon this foundation to establish with rigorous logic the science of numbers, and just in this consists the content of my wearisome labor which in all its essentials I had completed several years before the appearance of Cantor's memoir and at a time when the work of Bolzano was unknown to me even by name. For the benefit of those who are interested in and understand the difficulties of such an investigation, I add the following remark. We can lay down an entirely different definition of the finite and infinite, which appears still simpler since the notion of similarity of transformation is not even assumed, viz.:

"A system S is said to be finite when it may be so transformed in itself (36) that no proper part (6) of S is transformed in itself; in the contrary case S is called an infinite system."

Now let us attempt to erect our edifice upon this new foundation! We shall soon meet with serious difficulties, and I believe myself warranted in saying that the proof of the perfect agreement of this definition with the former can be obtained only (and then easily) when we are permitted to assume the series of natural numbers as already developed and to make use of the final considerations in (131); and yet nothing is said of all these things in either the one definition or the other! From this we can see how very great is the number of steps in thought needed for such a remodeling of a definition.

About a year after the publication of my memoir I became acquainted with G. Frege's *Grundlagen der Arithmetik*, which had already appeared in the year 1884. However different the view of the essence of number adopted in that work is from my own, yet it contains, particularly from § 79 on, points of very close contact with my paper, especially with my definition (44). The agreement, to be sure, is not easy to discover on account of the different form of expression; but the positiveness with which the author speaks of the logical inference from n to $n+1$ (page 93, below) shows plainly that here he stands upon the same ground with me. In the meantime E. Schröder's *Vorlesungen über die Algebra der Logik* has been almost completed (1890–1891). Upon the importance of this extremely suggestive work, to which I pay my highest tribute, it is impossible here to enter further; I will simply confess that in spite of the remark made on p. 253 of Part I., I have retained my somewhat clumsy symbols (8) and (17); they make no claim to be adopted generally but are intended simply to serve the purpose of this arithmetic paper to which in my view they are better adapted than sum and product symbols.

R. Dedekind.

Harzburg, August 24, 1893.

THE NATURE AND MEANING OF NUMBERS.

I.

SYSTEMS OF ELEMENTS.

1. In what follows I understand by *thing* every object of our thought. In order to be able easily to speak of things, we designate them by symbols, e. g., by letters, and we venture to speak briefly of the thing a or of a simply, when we mean the thing denoted by a and not at all the letter a itself. A thing is completely determined by all that can be affirmed or thought concerning it. A thing a is the same as b (identical with b), and b the same as a, when all that can be thought concerning a can also be thought concerning b, and when all that is true of b can also be thought of a. That a and b are only symbols or names for one and the same thing is indicated by the notation $a=b$, and also by $b=a$. If further $b=c$, that is, if c as well as a is a symbol for the thing denoted by b, then is also $a=c$. If the above coincidence of the thing denoted by a with the thing denoted by b does not exist, then are the things a, b said to be different, a is another thing than b, b another thing than

a; there is some property belonging to the one that does not belong to the other.

2. It very frequently happens that different things, *a*, *b*, *c*, . . . for some reason can be considered from a common point of view, can be associated in the mind, and we say that they form a *system* S; we call the things *a*, *b*, *c*, . . . *elements* of the system S, they are *contained* in S; conversely, S *consists* of these elements. Such a system S (an aggregate, a manifold, a totality) as an object of our thought is likewise a thing (1); it is completely determined when with respect to every thing it is determined whether it is an element of S or not.* The system S is hence the same as the system T, in symbols $S = T$, when every element of S is also element of T, and every element of T is also element of S. For uniformity of expression it is advantageous to include also the special case where a system S consists of a *single* (one and only one) element *a*, i. e., the thing *a* is element of S, but every thing different from *a* is not an element of S. On the other hand, we intend here for certain reasons wholly to exclude the empty system which contains no element at all, although for other

*In what manner this determination is brought about, and whether we know a way of deciding upon it, is a matter of indifference for all that follows; the general laws to be developed in no way depend upon it; they hold under all circumstances. I mention this expressly because Kronecker not long ago (*Crelle's Journal*, Vol. 99, pp. 334-336) has endeavored to impose certain limitations upon the free formation of concepts in mathematics which I do not believe to be justified; but there seems to be no call to enter upon this matter with more detail until the distinguished mathematician shall have published his reasons for the necessity or merely the expediency of these limitations.

investigations it may be appropriate to imagine such a system.

3. Definition. A system A is said to be *part* of a system S when every element of A is also element of S. Since this relation between a system A and a system S will occur continually in what follows, we shall express it briefly by the symbol $A \subseteq S$. The inverse symbol $S \supseteq A$, by which the same fact might be expressed, for simplicity and clearness I shall wholly avoid, but for lack of a better word I shall sometimes say that S is *whole* of A, by which I mean to express that among the elements of S are found all the elements of A. Since further every element s of a system S by (2) can be itself regarded as a system, we can hereafter employ the notation $s \subseteq S$.

4. Theorem. $A \subseteq A$, by reason of (3).

5. Theorem. If $A \subseteq B$ and $B \subseteq A$, then $A = B$.

The proof follows from (3), (2).

6. Definition. A system A is said to be a *proper* [*echter*] part of S, when A is part of S, but different from S. According to (5) then S is not a part of A, i. e., there is in S an element which is not an element of A.

7. Theorem. If $A \subseteq B$ and $B \subseteq C$, which may be denoted briefly by $A \subseteq B \subseteq C$, then is $A \subseteq C$, and A is certainly a proper part of C, if A is a proper part of B or if B is a proper part of C.

The proof follows from (3), (6).

8. Definition. By the system *compounded* out of

any systems A, B, C, . . . to be denoted by $\mathfrak{M}$ (A, B, C, . . .) we mean that system whose elements are determined by the following prescription: a thing is considered as element of $\mathfrak{M}$ (A, B, C, . . .) when and only when it is element of some one of the systems A, B, C, . . ., i. e., when it is element of A, *or* B, *or* C, . . . We include also the case where only a single system A exists; then obviously $\mathfrak{M}(A)=A$. We observe further that the system $\mathfrak{M}$ (A, B, C, . . .) compounded out of A, B, C, . . . is carefully to be distinguished from the system whose elements are the systems A, B, C, . . . themselves.

9. Theorem. The systems A, B, C, . . . are parts of $\mathfrak{M}$ (A, B, C, . . .).

The proof follows from (8), (3).

10. Theorem. If A, B, C, . . . are parts of a system S, then is $\mathfrak{M}(A, B, C, \ldots) \mathrel{3} S$.

The proof follows from (8), (3).

11. Theorem. If P is part of one of the systems A, B, C, . . . then is $P \mathrel{3} \mathfrak{M}(A, B, C, \ldots)$.

The proof follows from (9), (7).

12. Theorem. If each of the systems P, Q, . . . is part of one of the systems A, B, C, . . . then is $\mathfrak{M}(P, Q, \ldots) \mathrel{3} \mathfrak{M}(A, B, C, \ldots)$.

The proof follows from (11), (10).

13. Theorem. If A is compounded out of any of the systems P, Q, . . . then is $A \mathrel{3} \mathfrak{M}(P, Q, \ldots)$.

Proof. For every element of A is by (8) element of one of the systems P, Q, . . ., consequently by (8) also element of $\mathfrak{M}(P, Q, \ldots)$, whence the theorem follows by (3).

14. Theorem. If each of the systems $A, B, C, \ldots$ is compounded out of any of the systems $P, Q, \ldots$ then is

$$\mathfrak{M}(A, B, C, \ldots) \ni \mathfrak{M}(P, Q, \ldots)$$

The proof follows from (13), (10).

15. Theorem. If each of the systems $P, Q, \ldots$ is part of one of the systems $A, B, C, \ldots$, and if each of the latter is compounded out of any of the former, then is

$$\mathfrak{M}(P, Q, \ldots) = \mathfrak{M}(A, B, C, \ldots).$$

The proof follows from (12), (14), (5).

16. Theorem. If

$$A = \mathfrak{M}(P, Q) \text{ and } B = \mathfrak{M}(Q, R)$$
$$\text{then is } \mathfrak{M}(A, R) = \mathfrak{M}(P, B).$$

Proof. For by the preceding theorem (15)

$$\mathfrak{M}(A, R) \text{ as well as } \mathfrak{M}(P, B) = \mathfrak{M}(P, Q, R).$$

17. Definition. A thing g is said to be *common* element of the systems $A, B, C, \ldots$, if it is contained in each of these systems (that is in A *and* in B *and* in $C \ldots$). Likewise a system T is said to be a *common part* of $A, B, C, \ldots$ when T is part of each of these systems; and by the *community* [*Gemeinheit*] of the systems $A, B, C, \ldots$ we understand the perfectly determinate system $\mathfrak{G}(A, B, C, \ldots)$ which consists of all the common elements g of $A, B, C, \ldots$ and

hence is likewise a common part of those systems. We again include the case where only a single system A occurs; then $\mathfrak{G}(A)$ (is to be put) $= A$. But the case may also occur that the systems $A, B, C, \ldots$ possess no common element at all, therefore no common part, no community; they are then called systems *without* common part, and the symbol $\mathfrak{G}(A, B, C, \ldots)$ is meaningless (compare the end of (2)). We shall however almost always in theorems concerning communities leave it to the reader to add in thought the condition of their existence and to discover the proper interpretation of these theorems for the case of non-existence.

18. Theorem. Every common part of $A, B, C, \ldots$ is part of $\mathfrak{G}(A, B, C, \ldots)$.

The proof follows from (17).

19. Theorem. Every part of $\mathfrak{G}(A, B, C, \ldots)$ is common part of $A, B, C, \ldots$

The proof follows from (17), (7).

20. Theorem. If each of the systems $A, B, C, \ldots$ is whole (3) of one of the systems $P, Q, \ldots$ then is

$$\mathfrak{G}(P, Q, \ldots) \ni \mathfrak{G}(A, B, C, \ldots)$$

Proof. For every element of $\mathfrak{G}(P, Q, \ldots)$ is common element of $P, Q, \ldots$, therefore also common element of $A, B, C, \ldots$, which was to be proved.

II.

TRANSFORMATION OF A SYSTEM.

21. Definition.* By a *transformation* [*Abbildung*] ϕ of a system S we understand a law according to which to every determinate element s of S there *belongs* a determinate thing which is called the *transform* of s and denoted by $\phi(s)$; we say also that $\phi(s)$ *corresponds* to the element s, that $\phi(s)$ *results* or is *produced* from s by the transformation ϕ, that s is *transformed* into $\phi(s)$ by the transformation ϕ. If now T is any part of S, then in the transformation ϕ of S is likewise contained a determinate transformation of T, which for the sake of simplicity may be denoted by the same symbol ϕ and consists in this that to every element t of the system T there corresponds the same transform $\phi(t)$, which t possesses as element of S; at the same time the system consisting of all transforms $\phi(t)$ shall be called the transform of T and be denoted by $\phi(T)$, by which also the significance of $\phi(S)$ is defined. As an example of a transformation of a system we may regard the mere assignment of determinate symbols or names to its elements. The simplest transformation of a system is that by which each of its elements is transformed into itself; it will be called the *identical* transformation of the system. For convenience, in the following theorems (22), (23), (24), which deal with an arbitrary transformation ϕ of

*See Dirichlet's *Vorlesungen über Zahlentheorie*, 3d edition, 1879, § 163.

an arbitrary system S, we shall denote the transforms of elements s and parts T respectively by s' and T'; in addition we agree that small and capital italics without accent shall always signify elements and parts of this system S.

22. Theorem.* If $A \mathrel{3} B$, then $A' \mathrel{3} B'$.

Proof. For every element of A' is the transform of an element contained in A, and therefore also in B. and is therefore element of B', which was to be proved.

23. Theorem. The transform of $\mathfrak{M}\ (A, B, C, \ldots)$ is $\mathfrak{M}\ (A', B', C', \ldots)$.

Proof. If we denote the system $\mathfrak{M}\ (A, B, C, \ldots)$ which by (10) is likewise part of S by M, then is every element of its transform M' the transform m' of an element m of M; since therefore by (8) m is also element of one of the systems $A, B, C, \ldots$ and consequently m' element of one of the systems $A', B', C', \ldots$, and hence by (8) also element of $\mathfrak{M}\ (A', B', C', \ldots)$, we have by (3)

$$M' \mathrel{3} \mathfrak{M}(A', B', C', \ldots).$$

On the other hand, since $A, B, C, \ldots$ are by (9) parts of M, and hence $A', B', C', \ldots$ by (22) parts of M', we have by (10)

$$\mathfrak{M}(A', B', C', \ldots) \mathrel{3} M'.$$

By combination with the above we have by (5) the theorem to be proved

$$M' = \mathfrak{M}(A', B', C', \ldots).$$

*See theorem 27.

24. Theorem.* The transform of every common part of A, B, C, . . ., and therefore that of the community $\mathfrak{G}$ $(A, B, C, \ldots)$ is part of $\mathfrak{G}$ $(A', B', C', \ldots)$.

Proof. For by (22) it is common part of A', B', C', . . ., whence the theorem follows by (18).

25. Definition and theorem. If ϕ is a transformation of a system S, and ψ a transformation of the transform $S' = \phi(S)$, there always results a transformation θ of S, *compounded*† out of ϕ and ψ, which consists of this that to every element s of S there corresponds the transform

$$\theta(s) = \psi(s') = \psi(\phi(s)),$$

where again we have put $\phi(s) = s'$. This transformation θ can be denoted briefly by the symbol $\psi . \phi$ or $\psi \phi$, the transform $\theta(s)$ by $\psi \phi(s)$ where the order of the symbols ϕ, ψ is to be considered, since in general the symbol $\phi \psi$ has no interpretation and actually has meaning only when $\psi(s') 3 s$. If now χ signifies a transformation of the system $\psi(s') = \psi \phi(s)$ and η the transformation $\chi \psi$ of the system S' compounded out of ψ and χ, then is $\chi \theta(s) = \chi \psi(s') = \eta(s') = \eta \phi(s)$; therefore the compound transformations $\chi \theta$ and $\eta \phi$ coincide for every element s of S, i. e., $\chi \theta = \eta \phi$. In accordance with the meaning of θ and η this theorem can finally be expressed in the form

$$\chi \cdot \psi \phi = \chi \psi \cdot \phi,$$

*See theorem 29.

†A confusion of this compounding of transformations with that of systems of elements is hardly to be feared.

and this transformation compounded out of ϕ, ψ, χ can be denoted briefly by $\chi\psi\phi$.

III.

SIMILARITY OF A TRANSFORMATION. SIMILAR SYSTEMS.

26. Definition. A transformation ϕ of a system S is said to be *similar* [*ähnlich*] or *distinct*, when to different elements a, b of the system S there always correspond different transforms $a'=\phi(a)$, $b'=\phi(b)$. Since in this case conversely from $s'=t'$ we always have $s=t$, then is every element of the system $S'=\phi(S)$ the transform s' of a single, perfectly determinate element s of the system S, and we can therefore set over against the transformation ϕ of S an *inverse* transformation of the system S', to be denoted by $\bar{\phi}$, which consists in this that to every element s' of S' there corresponds the transform $\bar{\phi}(s')=s$, and obviously this transformation is also similar. It is clear that $\bar{\phi}(S')=S$, that further ϕ is the inverse transformation belonging to $\bar{\phi}$ and that the transformation $\bar{\phi}\phi$ compounded out of ϕ and $\bar{\phi}$ by (25) is the identical transformation of S (21). At once we have the following additions to II., retaining the notation there given.

27. Theorem.* If $A'\,3\,B'$, then $A\,3\,B$.

Proof. For if a is an element of A then is a' an element of A', therefore also of B', hence $=b'$, where b is an element of B; but since from $a'=b'$ we always

* See theorem 22.

have $a=b$, then is every element of A also element of B, which was to be proved.

28. Theorem. If $A'=B'$, then $A=B$.

The proof follows from (27), (4), (5).

29. Theorem.* If $G=\mathfrak{G}(A, B, C, \ldots)$, then

$$G'=\mathfrak{G}(A', B', C', \ldots).$$

Proof. Every element of $\mathfrak{G}(A', B', C', \ldots)$ is certainly contained in S', and is therefore the transform g' of an element g contained in S; but since g' is common element of $A', B', C', \ldots$ then by (27) must g be common element of $A, B, C, \ldots$ therefore also element of G; hence every element of $\mathfrak{G}(A', B', C', \ldots)$ is transform of an element g of G, therefore element of G', i. e., $\mathfrak{G}(A', B', C', \ldots) \ni G'$, and accordingly our theorem follows from (24), (5).

30. Theorem. The identical transformation of a system is always a similar transformation.

31. Theorem. If ϕ is a similar transformation of S and ψ a similar transformation of $\phi(S)$, then is the transformation $\psi\phi$ of S, compounded of ϕ and ψ, a similar transformation, and the associated inverse transformation $\overline{\psi\phi}=\overline{\phi}\,\overline{\psi}$.

Proof. For to different elements a, b of S correspond different transforms $a'=\phi(a)$, $b'=\phi(b)$, and to these again different transforms $\psi(a')=\psi\phi(a)$, $\psi(b')=\psi\phi(b)$ and therefore $\psi\phi$ is a similar transformation. Besides every element $\psi\phi(s)=\psi(s')$ of the system $\psi\phi(S)$ is transformed by $\overline{\psi}$ into $s'=\phi(s)$ and

* See theorem 24.

this by $\bar{\phi}$ into s, therefore $\psi\phi(s)$ is transformed by $\bar{\phi}\bar{\psi}$ into s, which was to be proved.

32. Definition. The systems R, S are said to be *similar* when there exists such a similar transformation ϕ of S that $\phi(S)=R$, and therefore $\bar{\phi}(R)=S$. Obviously by (30) every system is similar to itself.

33. Theorem. If R, S are similar systems, then every system Q similar to R is also similar to S.

Proof. For if ϕ, ψ are similar transformations of S, R such that $\phi(S)=R$, $\psi(R)=Q$, then by (31) $\psi\phi$ is a similar transformation of S such that $\psi\phi(S)=Q$, which was to be proved.

34. Definition. We can therefore separate all systems into *classes* by putting into a determinate class all systems $Q, R, S, \ldots$, and only those, that are similar to a determinate system R, the *representative* of the class; according to (33) the class is not changed by taking as representative any other system belonging to it.

35. Theorem. If R, S are similar systems, then is every part of S also similar to a part of R, every proper part of S also similar to a proper part of R.

Proof. For if ϕ is a similar transformation of S, $\phi(S)=R$, and $T\,3\,S$, then by (22) is the system similar to $T\,\phi(T)\,3\,R$; if further T is proper part of S, and s an element of S not contained in T, then by (27) the element $\phi(s)$ contained in R cannot be contained in $\phi(T)$; hence $\phi(T)$ is proper part of R, which was to be proved.

IV.

TRANSFORMATION OF A SYSTEM IN ITSELF.

36. Definition. If ϕ is a similar or dissimilar transformation of a system S, and $\phi(S)$ part of a system Z, then ϕ is said to be a transformation of S *in* Z, and we say S is transformed by ϕ in Z. Hence we call ϕ a transformation of the system S *in itself*, when $\phi(S) 3 S$, and we propose in this paragraph to investigate the general laws of such a transformation ϕ. In doing this we shall use the same notations as in II. and again put $\phi(s) = s'$, $\phi(T) = T'$. These transforms s', T' are by (22), (7) themselves again elements or parts of S, like all things designated by italic letters.

37. Definition. K is called a *chain* [*Kette*], when $K' 3 K$. We remark expressly that this name does not in itself belong to the part K of the system S, but is given only with respect to the particular transformation ϕ; with reference to another transformation of the system S in itself K can very well not be a chain.

38. Theorem. S is a chain.

39. Theorem. The transform K' of a chain K is a chain.

Proof. For from $K' 3 K$ it follows by (22) that $(K')' 3 K'$, which was to be proved.

40. Theorem. If A is part of a chain K, then is also $A' 3 K$.

Proof. For from $A\,3\,K$ it follows by (22) that $A'\,3\,K'$, and since by (37) $K'\,3\,K$, therefore by (7) $A'\,3\,K$, which was to be proved.

41. Theorem. If the transform A' is part of a chain L, then is there a chain K, which satisfies the conditions $A\,3\,K$, $K'\,3\,L$; and $\mathfrak{M}(A, L)$ is just such a chain K.

Proof. If we actually put $K = \mathfrak{M}(A, L)$, then by (9) the one condition $A\,3\,K$ is fulfilled. Since further by (23) $K' = \mathfrak{M}(A', L')$ and by hypothesis $A'\,3\,L$, $L'\,3\,L$, then by (10) is the other condition $K'\,3\,L$ also fulfilled and hence it follows because by (9) $L\,3\,K$, that also $K'\,3\,K$, i. e., K is a chain, which was to be proved.

42. Theorem. A system M compounded simply out of chains A, B, C, . . . is a chain.

Proof. Since by (23) $M' = \mathfrak{M}(A', B', C', \ldots)$ and by hypothesis $A'\,3\,B$, $B'\,3\,B$, $C'\,3\,C$, . . . therefore by (12) $M'\,3\,M$, which was to be proved.

43. Theorem. The community G of chains A B, C, . . . is a chain.

Proof. Since by (17) G is common part of A, B, C, . . ., therefore by (22) G' common part of A', B', C', . . ., and by hypothesis $A'\,3\,A$, $B'\,3\,B$, $C'\,3\,C$, . . ., then by (7) G' is also common part of A, B, C, . . . and hence by (18) also part of G, which was to be proved.

44. Definition. If A is any part of S, we will denote by A_0 the community of all those chains (e.g., S)

of which A is part; this community A_o exists (17) because A is itself common part of all these chains. Since further by (43) A_o is a chain, we will call A_o the *chain of the system* A, or briefly the chain of A. This definition too is strictly related to the fundamental determinate transformation ϕ of the system S in itself, and if later, for the sake of clearness, it is necessary we shall at pleasure use the symbol $\phi_o(A)$ instead of A_o, and likewise designate the chain of A corresponding to another transformation ω by $\omega_o(A)$. For this very important notion the following theorems hold true.

45. Theorem. $A \mathfrak{3} A_o$.

Proof. For A is common part of all those chains whose community is A_o, whence the theorem follows by (18).

46. Theorem. $(A_o)' \mathfrak{3} A_o$.

Proof. For by (44) A_o is a chain (37).

47. Theorem. If A is part of a chain K, then is also $A_o \mathfrak{3} K$.

Proof. For A_o is the community and hence also a common part of all the chains K, of which A is part.

48. Remark. One can easily convince himself that the notion of the chain A_o defined in (44) is completely characterised by the preceding theorems, (45), (46), (47).

49. Theorem. $A' \mathfrak{3} (A_o)'$.

The proof follows from (45), (22).

50. Theorem. $A' 3 A_0$.

The proof follows from (49), (46), (7).

51. Theorem. If A is a chain, then $A_0 = A$.

Proof. Since A is part of the chain A, then by (47) $A_0 3 A$, whence the theorem follows by (45), (5).

52. Theorem. If $B 3 A$, then $B 3 A_0$.

The proof follows from (45), (7).

53. Theorem. If $B 3 A_0$, then $B_0 3 A_0$, and conversely.

Proof. Because A_0 is a chain, then by (47) from $B 3 A_0$, we also get $B_0 3 A_0$; conversely, if $B_0 3 A_0$, then by (7) we also get $B 3 A_0$, because by (45) $B 3 B_0$.

54. Theorem. If $B 3 A$, then is $B_0 3 A_0$.

The proof follows from (52), (53).

55. Theorem. If $B 3 A_0$, then is also $B' 3 A_0$.

Proof. For by (53) $B_0 3 A_0$, and since by (50) $B' 3 B_0$, the theorem to be proved follows by (7). The same result, as is easily seen, can be obtained from (22), (46), (7), or also from (40).

56. Theorem. If $B 3 A_0$, then is $(B_0)' 3 (A_0)'$.

The proof follows from (53), (22).

57. Theorem and definition. $(A_0)' = (A')_0$, i. e., the transform of the chain of A is at the same time the chain of the transform of A. Hence we can designate this system in short by A'_0 and at pleasure call it the *chain-transform* or *transform-chain* of A. With the clearer notation given in (44) the theorem might be expressed by $\phi(\phi_0(A)) = \phi_0(\phi(A))$.

Proof. If for brevity we put $(A')_0 = L$, L is a

chain (44) and by (45) $A' \mathfrak{3} L$; hence by (41) there exists a chain K satisfying the conditions $A \mathfrak{3} K$, $K' \mathfrak{3} L$; hence from (47) we have $A_o \mathfrak{3} K$, therefore $(A_o)' \mathfrak{3} K'$, and hence by (7) also $(A_o)' \mathfrak{3} L$, i. e.,

$$(A_o)' \mathfrak{3} (A')_o.$$

Since further by (49) $A' \mathfrak{3} (A_o)'$, and by (44), (39) $(A_o)'$ is a chain, then by (47) also

$$(A')_o \mathfrak{3} (A_o)',$$

whence the theorem follows by combining with the preceding result (5).

58. Theorem. $A_o = \mathfrak{M}(A, A'_o)$, i. e., the chain of A is compounded out of A and the transform-chain of A.

Proof. If for brevity we again put

$$L = A'_o = (A_o)' = (A')_o \text{ and } K = \mathfrak{M}(A, L),$$

then by (45) $A' \mathfrak{3} L$, and since L is a chain, by (41) the same thing is true of K; since further $A \mathfrak{3} K$ (9), therefore by (47)

$$A_o \mathfrak{3} K.$$

On the other hand, since by (45) $A \mathfrak{3} A_o$, and by (46) also $L \mathfrak{3} A_o$, then by (10) also

$$K \mathfrak{3} A_o,$$

whence the theorem to be proved $A_o = K$ follows by combining with the preceding result (5).

59. Theorem of complete induction. In order to show that the chain A_o is part of any system Σ— be this latter part of S or not—it is sufficient to show,

ρ. that $A \mathfrak{3} \Sigma$, and

σ. that the transform of every common element of A_o and Σ is likewise element of Σ.

Proof. For if ρ is true, then by (45) the community $G = \mathfrak{G}(A_o, \Sigma)$ certainly exists, and by (18) $A \, 3 \, G$; since besides by (17)

$$G \, 3 \, A_o,$$

then is G also part of our system S, which by ϕ is transformed in itself and at once by (55) we have also $G' \, 3 \, A_o$. If then σ is likewise true, i. e., if $G' \, 3 \, \Sigma$, then must G' as common part of the systems A_o, Σ by (18) be part of their community G, i. e., G is a chain (37), and since, as above noted, $A \, 3 \, G$, then by (47) is also

$$A_o \, 3 \, G,$$

and therefore by combination with the preceding result $G = A_o$, hence by (17) also $A_o \, 3 \, \Sigma$, which was to be proved.

60. The preceding theorem, as will be shown later, forms the scientific basis for the form of demonstration known by the name of complete induction (the inference from n to $n+1$); it can also be stated in the following manner: In order to show that all elements of the chain A_o possess a certain property $\mathfrak{E}$ (or that a theorem $\mathfrak{S}$ dealing with an undetermined thing n actually holds good for all elements n of the chain A_o) it is sufficient to show

ρ. that all elements a of the system A possess the property $\mathfrak{E}$ (or that $\mathfrak{S}$ holds for all a's) and

σ. that to the transform n' of every such element n of A_o possessing the property $\mathfrak{E}$, belongs the same property $\mathfrak{E}$ (or that the theorem $\mathfrak{S}$, as soon as it holds for an element n of A_o, certainly must also hold for its transform n').

Indeed, if we denote by Σ the system of all things possessing the property $\mathfrak{E}$ (or for which the theorem $\mathfrak{S}$ holds) the complete agreement of the present manner of stating the theorem with that employed in (59) is immediately obvious.

61. Theorem. The chain of $\mathfrak{M}(A, B, C, \ldots)$ is $\mathfrak{M}(A_o, B_o, C_o, \ldots)$.

Proof. If we designate by M the former, by K the latter system, then by (42) K is a chain. Since then by (45) each of the systems $A, B, C, \ldots$ is part of one of the systems $A_o, B_6, C_o, \ldots$, and therefore by (12) $M\,3\,K$, then by (47) we also have

$$M_o\,3\,K.$$

On the other hand, since by (9) each of the systems $A, B, C, \ldots$ is part of M, and hence by (45), (7) also part of the chain M_o, then by (47) must also each of the systems $A_o, B_o, C_o, \ldots$ be part of M_o, therefore by (10)

$$K\,3\,M_o$$

whence by combination with the preceding result follows the theorem to be proved $M_o = K$ (5).

62. Theorem. The chain of $\mathfrak{G}(A, B, C, \ldots)$ is part of $\mathfrak{G}(A_o, B_o, C_o, \ldots)$.

Proof. If we designate by G the former, by K the latter system, then by (43) K is a chain. Since then each of the systems $A_o, B_o, C_o, \ldots$ by (45) is whole of one of the systems A, B, C, . . ., and hence by (20) $G \mathrel{3} K$, therefore by (47) we obtain the theorem to be proved $G_o \mathrel{3} K$.

63. Theorem. If $K' \mathrel{3} L \mathrel{3} K$, and therefore K is a chain, L is also a chain. If the same is proper part of K, and U the system of all those elements of K which are not contained in L, and if further the chain U_o is proper part of K, and V the system of all those elements of K which are not contained in U_o, then is $K = \mathfrak{M}(U_o, V)$ and $L = \mathfrak{M}(U'_o, V)$. If finally $L = K'$ then $V \mathrel{3} V'$.

The proof of this theorem of which (as of the two preceding) we shall make no use may be left for the reader.

V.

THE FINITE AND INFINITE.

64. Definition.* A system S is said to be *infinite* when it is similar to a proper part of itself (32); in the contrary case S is said to be a *finite* system.

65. Theorem. Every system consisting of a single element is finite.

Proof. For such a system possesses no proper part (2), (6).

*If one does not care to employ the notion of similar systems (32) he must say: S is said to be infinite, when there is a proper part of S (6) in which S can be distinctly (similarly) transformed (26), (36). In this form I submitted the definition of the infinite which forms the core of my whole investigation in September, 1882, to G. Cantor and several years earlier to Schwarz and Weber. All other attempts that have come to my knowledge to distinguish the infinite from the finite seem to me to have met with so little success that I think I may be permitted to forego any criticism of them.

66. Theorem. There exist infinite systems.

Proof.* My own realm of thoughts, i. e., the totality S of all things, which can be objects of my thought, is infinite. For if s signifies an element of S, then is the thought s', that s can be object of my thought, itself an element of S. If we regard this as transform $\phi(s)$ of the element s then has the transformation ϕ of S, thus determined, the property that the transform S' is part of S; and S' is certainly proper part of S, because there are elements in S (e. g., my own ego) which are different from such thought s' and therefore are not contained in S'. Finally it is clear that if a, b are different elements of S, their transforms a', b' are also different, that therefore the transformation ϕ is a distinct (similar) transformation (26). Hence S is infinite, which was to be proved.

67. Theorem. If R, S are similar systems, then is R finite or infinite according as S is finite or infinite.

Proof. If S is infinite, therefore similar to a proper part S' of itself, then if R and S are similar, S' by (33) must be similar to R and by (35) likewise similar to a proper part of R, which therefore by (33) is itself similar to R; therefore R is infinite, which was to be proved.

68. Theorem. Every system S, which possesses an infinite part is likewise infinite; or, in other words, every part of a finite system is finite.

*A similar consideration is found in § 13 of the *Paradoxien des Unendlichen* by Bolzano (Leipzig, 1851).

Proof. If T is infinite and there is hence such a similar transformation ψ of T, that $\psi(T)$ is a proper part of T, then, if T is part of S, we can extend this transformation ψ to a transformation ϕ of S in which, if s denotes any element of S, we put $\phi(s)=\psi(s)$ or $\phi(s)=s$ according as s is element of T or not. This transformation ϕ is a similar one; for, if a, b denote different elements of S, then if both are contained in T, the transform $\phi(a)=\psi(a)$ is different from the transform $\phi(b)=\psi(b)$, because ψ is a similar transformation; if further a is contained in T, but b not, then is $\phi(a)=\psi(a)$ different from $\phi(b)=b$, because $\psi(a)$ is contained in T; if finally neither a nor b is contained in T then also is $\phi(a)=a$ different from $\phi(b)=b$, which was to be shown. Since further $\psi(T)$ is part of T, because by (7) also part of S, it is clear that also $\phi(S) 3 S$. Since finally $\psi(T)$ is proper part of T there exists in T and therefore also in S, an element t, not contained in $\psi(T)=\phi(T)$; since then the transform $\phi(s)$ of every element s not contained in T is equal to s, and hence is different from t, t cannot be contained in $\phi(S)$; hence $\phi(S)$ is proper part of S and consequently S is infinite, which was to be proved.

69. Theorem. Every system which is similar to a part of a finite system, is itself finite.

The proof follows from (67), (68).

70. Theorem. If a is an element of S, and if the aggregate T of all the elements of S different from a is finite, then is also S finite.

Proof. We have by (64) to show that if ϕ denotes any similar transformation of S in itself, the transform $\phi(S)$ or S' is never a proper part of S but always $=S$. Obviously $S=\mathfrak{M}(a, T)$ and hence by (23), if the transforms are again denoted by accents, $S'=\mathfrak{M}(a', T')$, and, on account of the similarity of the transformation ϕ, a' is not contained in T' (26). Since further by hypothesis $S' \mathrel{3} S$, then must a' and likewise every element of T' either $=a$, or be element of T. If then—a case which we will treat first—a is not contained in T', then must $T' \mathrel{3} T$ and hence $T'=T$, because ϕ is a similar transformation and because T is a finite system; and since a', as remarked, is not contained in T', i.e., not in T, then must $a'=a$, and hence in this case we actually have $S'=S$ as was stated. In the opposite case when a is contained in T' and hence is the transform b' of an element b contained in T, we will denote by U the aggregate of all those elements u of T, which are different from b; then $T=\mathfrak{M}(b, U)$ and by (15) $S=\mathfrak{M}(a, b, U)$, hence $S'=\mathfrak{M}(a', a, U')$. We now determine a new transformation ψ of T in which we put $\psi(b)=a'$, and generally $\psi(u)=u'$, whence by (23) $\psi(T)=\mathfrak{M}(a', U')$. Obviously ψ is a similar transformation, because ϕ was such, and because a is not contained in U and therefore also a' not in U'. Since further a and every element u is different from b then (on account of the similarity of ϕ) must also a' and every element u' be different from a and consequently contained in T; hence $\psi(T) \mathrel{3} T$

and since T is finite, therefore must $\psi(T)=T$, and $\mathfrak{M}(a', U')=T$. From this by (15) we obtain

$$\mathfrak{M}(a', a, U')=\mathfrak{M}(a, T)$$

i. e., according to the preceding $S'=S$. Therefore in this case also the proof demanded has been secured.

VI.

SIMPLY INFINITE SYSTEMS. SERIES OF NATURAL NUMBERS.

71. Definition. A system N is said to be *simply infinite* when there exists a similar transformation ϕ of N in itself such that N appears as chain (44) of an element not contained in $\phi(N)$. We call this element, which we shall denote in what follows by the symbol 1, the *base-element* of N and say the simply infinite system N is *set in order* [*geordnet*] by this transformation ϕ. If we retain the earlier convenient symbols for transforms and chains (IV) then the essence of a simply infinite system N consists in the existence of a transformation ϕ of N and an element 1 which satisfy the following conditions α, β, γ, δ:

α. $N' \mathrel{3} N$.

β. $N=1_o$.

γ. The element 1 is not contained in N'.

δ. The transformation ϕ is similar.

Obviously it follows from α, γ, δ that every simply infinite system N is actually an infinite system (64) because it is similar to a proper part N' of itself.

. Theorem. In every infinite system S a simply infinite system N is contained as a part.

Proof. By (64) there exists a similar transformation ϕ of S such that $\phi(S)$ or S' is a proper part of S; hence there exists an element 1 in S which is not contained in S'. The chain $N = 1_0$, which corresponds to this transformation ϕ of the system S in itself (44), is a simply infinite system set in order by ϕ; for the characteristic conditions α, β, γ, δ in (71) are obviously all fulfilled.

73. Definition. If in the consideration of a simply infinite system N set in order by a transformation ϕ we entirely neglect the special character of the elements; simply retaining their distinguishability and taking into account only the relations to one another in which they are placed by the order-setting transformation ϕ, then are these elements called *natural numbers* or *ordinal numbers* or simply *numbers*, and the base-element 1 is called the *base-number* of the *number-series* N. With reference to this freeing the elements from every other content (abstraction) we are justified in calling numbers a free creation of the human mind. The relations or laws which are derived entirely from the conditions α, β, γ, δ in (71) and therefore are always the same in all ordered simply infinite systems, whatever names may happen to be given to the individual elements (compare 134), form the first object of the *science of numbers* or *arithmetic*. From the general notions and theorems of IV. about the transformation

of a system in itself we obtain immediately the following fundamental laws where $a, b, \ldots m, n, \ldots$ always denote elements of N, therefore numbers, $A, B, C, \ldots$ parts of N, $a', b', \ldots m', n', \ldots A', B', C' \ldots$ the corresponding transforms, which are produced by the order-setting transformation ϕ and are always elements or parts of N; the transform n' of a number n is also called the number *following* n.

74. Theorem. Every number n by (45) is contained in its chain n_o and by (53) the condition $n 3 m_o$ is equivalent to $n_o 3 m_o$.

75. Theorem. By (57) $n'_o = (n_o)' = (n')_o$.

76. Theorem. By (46) $n'_o 3 n_o$.

77. Theorem. By (58) $n_o = \mathfrak{M}(n, n'_o)$.

78. Theorem. $N = \mathfrak{M}(1, N')$, hence every number different from the base-number 1 is element of N', i. e., transform of a number.

The proof follows from (77) and (71).

79. Theorem. N is the only number-chain containing the base-number 1.

Proof. For if 1 is element of a number-chain K, then by (47) the associated chain $N 3 K$, hence $N = K$, because it is self-evident that $K 3 N$.

80. Theorem of complete induction (inference from n to n'). In order to show that a theorem holds for all numbers n of a chain m_o, it is sufficient to show,

ρ. that it holds for $n = m$, and

σ. that from the validity of the theorem for a number n of the chain m_o its validity for the following number n' always follows.

This results immediately from the more general theorem (59) or (60). The most frequently occurring case is where $m=1$ and therefore m_o is the complete number-series N.

VII.

GREATER AND LESS NUMBERS.

81. Theorem. Every number n is different from the following number n'.

Proof by complete induction (80):

ρ. The theorem is true for the number $n=1$, because it is not contained in N' (71), while the following number $1'$ as transform of the number 1 contained in N is element of N'.

σ. If the theorem is true for a number n and we put the following number $n'=p$, then is n different from p, whence by (26) on account of the similarity (71) of the order-setting transformation ϕ it follows that n', and therefore p, is different from p'. Hence the theorem holds also for the number p following n, which was to be proved.

82. Theorem. In the transform-chain n'_o of a number n by (74), (75) is contained its transform n', but not the number n itself.

Proof by complete induction (80):

ρ. The theorem is true for $n=1$, because $1'_o=N'$,

and because by (71) the base-number 1 is not contained in N'.

σ. If the theorem is true for a number n, and we again put $n' = p$, then is n not contained in p_o, therefore is it different from every number q contained in p_o, whence by reason of the similarity of ϕ it follows that n', and therefore p, is different from every number q' contained in p'_o, and is hence not contained in p'_o. Therefore the theorem holds also for the number p following n, which was to be proved.

83. Theorem. The transform-chain n'_o is proper part of the chain n_o.

The proof follows from (76), (74), (82).

84. Theorem. From $m_o = n_o$ it follows that $m = n$.

Proof. Since by (74) m is contained in m_o, and

$$m_o = n_o = \mathfrak{M}(n, n'_o)$$

by (77), then if the theorem were false and hence m different from n, m would be contained in the chain n'_o, hence by (74) also $m_o 3 n'_o$, i. e., $n_o 3 n'_o$; but this contradicts theorem (83). Hence our theorem is established.

85. Theorem. If the number n is not contained in the number-chain K, then is $K 3 n'_o$.

Proof by complete induction (80):

ρ. By (78) the theorem is true for $n = 1$.

σ. If the theorem is true for a number n, then is it also true for the following number $p = n'$; for if p is not contained in the number-chain K, then by (40) n also cannot be contained in K and hence by our hypothesis $K \mathfrak{3} n'_0$; now since by (77) $n'_0 = p_0 = \mathfrak{M}(p, p'_0)$, hence $K \mathfrak{3} \mathfrak{M}(p, p'_0)$ and p is not contained in K, then must $K \mathfrak{3} p'_0$, which was to be proved.

86. Theorem. If the number n is not contained in the number-chain K, but its transform n' is, then $K = n'_0$.

Proof. Since n is not contained in K, then by (85) $K \mathfrak{3} n'_0$, and since $n' \mathfrak{3} K$, then by (47) is also $n'_0 \mathfrak{3} K$, and hence $K = n'_0$, which was to be proved.

87. Theorem. In every number-chain K there exists one, and by (84) only one, number k, whose chain $k_0 = K$.

Proof. If the base-number 1 is contained in K, then by (79) $K = N = 1_0$. In the opposite case let Z be the system of all numbers not contained in K; since the base-number 1 is contained in Z, but Z is only a proper part of the number-series N, then by (79) Z cannot be a chain, i. e., Z' cannot be part of Z; hence there exists in Z a number n, whose transform n' is not contained in Z, and is therefore certainly contained in K; since further n is contained in Z, and therefore not in K, then by (86) $K = n'_0$, and hence $k = n'$, which was to be proved.

88. Theorem. If m, n are different numbers then by (83), (84) one and only one of the chains m_0, n_0 is proper part of the other and either $n_0 \mathfrak{3} m'_0$ or $m_0 \mathfrak{3} n'_0$.

Proof. If n is contained in m_0, and hence by (74) also $n_0 \mathfrak{3} m_0$, then m can not be contained in the chain n_0 (because otherwise by (74) we should have $m_0 \mathfrak{3} n_0$, therefore $m_o = n_o$, and hence by (84) also $m = n$) and thence it follows by (85) that $n_o 3 m'_o$. In the contrary case, when n is not contained in the chain m_o, we must have by (85) $m_o 3 n'_o$, which was to be proved.

89. Definition. The number m is said to be *less* than the number n and at the same time n *greater* than m, in symbols

$$m < n, \quad n > m,$$

when the condition

$$n_o 3 m'_o$$

is fulfilled, which by (74) may also be expressed

$$n 3 m'_o.$$

90. Theorem. If m, n are any numbers, then always one and only one of the following cases λ, μ, ν occurs:

λ. $m = n$, $n = m$, i. e., $m_o = n_o$

μ. $m < n$, $n > m$, i. e., $n_o 3 m'_o$

ν. $m > n$, $n < m$, i. e., $m_o 3 n'_o$.

Proof. For if λ occurs (84) then can neither μ nor ν occur because by (83) we never have $n_o 3 n'_o$. But if λ does not occur then by (88) one and only one of the cases μ, ν occurs, which was to be proved.

91. Theorem. $n < n'$.

Proof. For the condition for the case ν in (90) is fulfilled by $m = n'$.

92. Definition. To express that m is either $= n$ or $< n$, hence not $> n$ (90) we use the symbols

$$m \leqq n \text{ or also } n \geqq m$$

and we say m is *at most equal* to n, and n is *at least equal* to m.

93. Theorem. Each of the conditions

$$m \leqq n,\quad m < n',\quad n_o 3 m_o$$

is equivalent to each of the others.

Proof. For if $m \leqq n$, then from λ, μ in (90) we always have $n_o 3 m_o$, because by (76) $m'_o 3 m$. Conversely, if $n_o 3 m_o$, and therefore by (74) also $n 3 m_o$, it follows from $m_o = \mathfrak{M}(m, m'_o)$ that either $n = m$, or $n 3 m'_o$, i. e., $n > m$. Hence the condition $m \leqq n$ is equivalent to $n_o 3 m_o$. Besides it follows from (22), (27), (75) that this condition $n_o 3 m_o$ is again equivalent to $n'_o 3 m'_o$, i. e., by μ in (90) to $m < n'$, which was to be proved.

94. Theorem. Each of the conditions

$$m' \leqq n,\quad m' < n',\quad m < n$$

is equivalent to each of the others.

The proof follows immediately from (93), if we replace in it m by m', and from μ in (90).

95. Theorem. If $l < m$ and $m \leqq n$ or if $l \leqq m$, and $m < n$, then is $l < n$. But if $l \leqq m$ and $m \leqq n$, then is $l \leqq n$.

Proof. For from the corresponding conditions (89), (93) $m_o 3 l'_o$ and $n_o 3 m_o$, we have by (7) $n_o 3 l'_o$ and the same thing comes also from the conditions $m_o 3 l_o$ and $n_o 3 m'_o$, because in consequence of the former we have also $m'_o 3 l'_o$. Finally from $m_o 3 l_o$ and $n_o 3 m_o$ we have also $n_o 3 l_o$, which was to be proved.

96. Theorem. In every part T of N there exists one and only one *least* number k, i. e., a number k

which is less than every other number contained in T. If T consists of a single number, then is it also the least number in T.

Proof. Since T_0 is a chain (44), then by (87) there exists one number k whose chain $k_0 = T_0$. Since from this it follows by (45), (77) that $T \mathfrak{Z} \mathfrak{M}(k, k'_0)$, then first must k itself be contained in T (because otherwise $T \mathfrak{Z} k'_0$, hence by (47) also $T_0 \mathfrak{Z} k'_0$, i. e., $k \mathfrak{Z} k'_0$, which by (83) is impossible), and besides every number of the system T, different from k, must be contained in k'_0, i. e., be $> k$ (89), whence at once from (90) it follows that there exists in T one and only one least number, which was to be proved.

97. Theorem. The least number of the chain n_0 is n, and the base-number 1 is the least of all numbers.

Proof. For by (74), (93) the condition $m \mathfrak{Z} n_0$ is equivalent to $m \geqq n$. Or our theorem also follows immediately from the proof of the preceding theorem, because if in that we assume $T = n_0$, evidently $k = n$ (51).

98. Definition. If n is any number, then will we denote by Z_n the system of all numbers that are *not greater* than n, and hence *not* contained in n'_0. The condition

$$m \mathfrak{Z} Z_n$$

by (92), (93) is obviously equivalent to each of the following conditions:

$$m \leqq n, \quad m < n', \quad n_0 \mathfrak{Z} m_0.$$

99. Theorem. $1 \mathfrak{Z} Z_n$ and $n \mathfrak{Z} Z_n$.

The proof follows from (98) or from (71) and (82).

100. Theorem. Each of the conditions equivalent by (98)

$$m \ni Z_n,\quad m \leqq n,\quad m < n',\quad n_o \ni m_o$$

is also equivalent to the condition

$$Z_m \ni Z_n.$$

Proof. For if $m \ni Z_n$, and hence $m \leqq n$, and if $l \ni Z_m$, and hence $l \leqq m$, then by (95) also $l \leqq n$, i. e., $l \ni Z_n$; if therefore $m \ni Z_n$, then is every element l of the system Z_m also element of Z_n, i. e., $Z_m \ni Z_n$. Conversely, if $Z_m \ni Z_n$, then by (7) must also $m \ni Z_n$, because by (99) $m \ni Z_m$, which was to be proved.

101. Theorem. The conditions for the cases λ, μ, ν in (90) may also be put in the following form:

$$\lambda.\quad m = n,\quad n = m,\quad Z_m = Z_n$$

$$\mu.\quad m < n,\quad n > m,\quad Z_{m'} \ni Z_n$$

$$\nu.\quad m > n,\quad n < m,\quad Z_{n'} \ni Z_m.$$

The proof follows immediately from (90) if we observe that by (100) the conditions $n_o \ni m_o$ and $Z_m \ni Z_n$ are equivalent.

102. Theorem. $Z_1 = 1$.

Proof. For by (99) the base-number 1 is contained in Z_1, while by (78) every number different from 1 is contained in $1'_o$, hence by (98) not in Z_1, which was to be proved.

103. Theorem. By (98) $N = \mathfrak{M}(Z_n, n'_o)$.

104. Theorem. $n = \mathfrak{G}(Z_n, n_o)$, i. e., n is the only common element of the system Z_n and n_o.

Proof. From (99) and (74) it follows that n is

contained in Z_n and n_o; but every element of the chain n_o different from n by (77) is contained in n'_o, and hence by (98) not in Z_n, which was to be proved.

105. Theorem. By (91), (98) the number n' is not contained in Z_n.

106. Theorem. If $m < n$, then is Z_m proper part of Z_n and conversely.

Proof. If $m < n$, then by (100) $Z_m \mathfrak{3} Z_n$, and since the number n, by (99) contained in Z_n, can by (98) not be contained in Z_m because $n > m$, therefore Z_m is proper part of Z_n. Conversely if Z_m is proper part of Z_n then by (100) $m \leqq n$, and since m cannot be $= n$, because otherwise $Z_m = Z_n$, we must have $m < n$, which was to be proved.

107. Theorem. Z_n is proper part of $Z_{n'}$.

The proof follows from (106), because by (91) $n < n'$.

108. Theorem. $Z_{n'} = \mathfrak{M}(Z_n, n')$.

Proof. For every number contained in $Z_{n'}$ by (98) is $\leqq n'$, hence either $= n'$ or $< n'$, and therefore by (98) element of Z_n. Therefore certainly $Z_{n'} \mathfrak{3} \mathfrak{M}(Z_n, n')$. Since conversely by (107) $Z_n \mathfrak{3} Z_{n'}$ and by (99) $n' \mathfrak{3} Z_{n'}$, then by (10) we have

$$\mathfrak{M}(Z_n, n') \mathfrak{3} Z_{n'},$$

whence our theorem follows by (5).

109. Theorem. The transform Z'_n of the system Z_n is proper part of the system $Z_{n'}$.

Proof. For every number contained in Z'_n is the transform m' of a number m contained in Z_n, and since

$m \leqq n$, and hence by (94) $m' \leqq n'$, we have by (98) $Z'_n 3 Z_{n'}$. Since further the number 1 by (99) is contained in $Z_{n'}$, but by (71) is not contained in the transform Z'_n, then is Z'_n proper part of $Z_{n'}$, which was to be proved.

110. Theorem. $Z_{n'} = \mathfrak{M}(1, Z'_n)$.

Proof. Every number of the system $Z_{n'}$ different from 1 by (78) is the transform m' of a number m and this must be $\leqq n$, and hence by (98) contained in Z_n (because otherwise $m > n$, hence by (94) also $m' > n'$ and consequently by (98) m' would not be contained in $Z_{n'}$); but from $m 3 Z_n$ we have $m' 3 Z'_n$, and hence certainly

$$Z_{n'} 3 \mathfrak{M}(1, Z'_n).$$

Since conversely by (99) $1 3 Z_{n'}$, and by (109) $Z'_n 3 Z_{n'}$, then by (10) we have $\mathfrak{M}(1, Z'_n) 3 Z_{n'}$ and hence our theorem follows by (5).

111. Definition. If in a system E of numbers there exists an element g, which is greater than every other number contained in E, then g is said to be the *greatest* number of the system E, and by (90) there can evidently be only one such greatest number in E. If a system consists of a single number, then is this number itself the greatest number of the system.

112. Theorem. By (98) n is the greatest number of the system Z_n.

113. Theorem. If there exists in E a greatest number g, then is $E 3 Z_g$.

Proof. For every number contained in E is $\leqq g$,

and hence by (98) contained in Z_{g}, which was to be proved.

114. Theorem. If E is part of a system Z_n, or what amounts to the same thing, there exists a number n such that all numbers contained in E are $\leqq n$, then E possesses a greatest number g.

Proof. The system of all numbers p satisfying the condition $E \mathfrak{3} Z_p$—and by our hypothesis such numbers exist—is a chain (37), because by (107), (7) it follows also that $E \mathfrak{3} Z_{p'}$, and hence by (87) $= g_o$, where g signifies the least of these numbers (96), (97). Hence also $E \mathfrak{3} Z_g$, therefore by (98) every number contained in E is $\leqq g$, and we have only to show that the number g is itself contained in E. This is immediately obvious if $g = 1$, for then by (102) Z_g, and consequently also E consists of the single number 1. But if g is different from 1 and consequently by (78) the transform f' of a number f, then by (108) is $E \mathfrak{3} \mathfrak{M}(Z_f, g)$; if therefore g were not contained in E, then would $E \mathfrak{3} Z_f$, and there would consequently be among the numbers p a number f by (91) $< g$, which is contrary to what precedes; hence g is contained in E, which was to be proved.

115. Definition. If $l < m$ and $m < n$ we say the number m *lies between* l and n (also between n and l).

116. Theorem. There exists no number lying between n and n'.

Proof. For as soon as $m < n'$, and hence by (93)

$m \leqq n$, then by (90) we cannot have $n < m$, which was to be proved.

117. Theorem. If t is a number in T, but not the least (96), then there exists in T one and only one *next less* number s, i. e., a number s such that $s < t$, and that there exists in T no number lying between s and t. Similarly, if t is not the greatest number in T (111) there always exists in T one and only one *next greater* number u, i. e., a number u such that $t < u$, and that there exists in T no number lying between t and u. At the same time in T t is next greater than s and next less than u.

Proof. If t is not the least number in T, then let E be the system of all those numbers of T that are $< t$; then by (98) $E \, 3 \, Z_t$, and hence by (114) there exists in E a greatest number s obviously possessing the properties stated in the theorem, and also it is the only such number. If further t is not the greatest number in T, then by (96) there certainly exists among all the numbers of T, that are $> t$, a least number u, which and which alone possesses the properties stated in the theorem. In like manner the correctness of the last part of the theorem is obvious.

118. Theorem. In N the number n' is next greater than n, and n next less than n'.

The proof follows from (116), (117).

VIII.

FINITE AND INFINITE PARTS OF THE NUMBER-SERIES.

119. Theorem. Every system Z_n in (98) is finite.

Proof by complete induction (80).

ρ. By (65), (102) the theorem is true for $n=1$.

σ. If Z_n is finite, then from (108) and (70) it follows that $Z_{n'}$ is also finite, which was to be proved.

120. Theorem. If m, n are different numbers, then are Z_m, Z_n dissimilar systems.

Proof. By reason of the symmetry we may by (90) assume that $m < n$; then by (106) Z_m is proper part of Z_n, and since by (119) Z_n is finite, then by (64) Z_m and Z_n cannot be similar, which was to be proved.

121. Theorem. Every part E of the number-series N, which possesses a greatest number (111), is finite.

The proof follows from (113), (119), (68).

122. Theorem. Every part U of the number-series N, which possesses no greatest number, is simply infinite (71).

Proof. If u is any number in U, there exists in U by (117) one and only one next greater number than u, which we will denote by $\psi(u)$ and regard as transform of u. The thus perfectly determined transformation ψ of the system U has obviously the property

$$\alpha. \quad \psi(U) \mathrel{3} U,$$

i. e., U is transformed in itself by ψ. If further u, v

are different numbers in U, then by symmetry we may by (90) assume that $u < v$; thus by (117) it follows from the definition of ψ that $\psi(u) \leqq v$ and $v < \psi(v)$, and hence by (95) $\psi(u) < \psi(v)$; therefore by (90) the transforms $\psi(u)$, $\psi(v)$ are different, i. e.,

δ. the transformation ψ is similar.

Further, if u_1 denotes the least number (96) of the system U, then every number u contained in U is $\geqq u_1$, and since generally $u < \psi(u)$, then by (95) $u_1 < \psi(u)$, and therefore by (90) u_1 is different from $\psi(u)$, i. e.,

γ. the element u_1 of U is not contained in $\psi(U)$.

Therefore $\psi(U)$ is proper part of U and hence by (64) U is an infinite system. If then in agreement with (44) we denote by $\psi_o(V)$, when V is any part of U, the chain of V corresponding to the transformation ψ, we wish to show finally that

$$\beta. \quad U = \psi_o(u_1).$$

In fact, since every such chain $\psi_o(V)$ by reason of its definition (44) is a part of the system U transformed in itself by ψ, then evidently is $\psi_o(u_1) \, 3 \, U$; conversely it is first of all obvious from (45) that the element u_1 contained in U is certainly contained in $\psi_o(u_1)$; but if we assume that there exist elements of U, that are not contained in $\psi_o(u_1)$, then must there be among them by (96) a least number w, and since by what precedes this is different from the least number u_1 of the system U, then by (117) must there exist in U also a number v which is next less than w, whence it follows at once that $w = \phi(v)$; since therefore $v < w$, then must v by reason of the definition of w certainly be contained in $\psi_o(u_1)$; but from this by (55) it follows that also $\psi(v)$, and hence w must be contained in $\psi_o(u_1)$, and since this is contrary to the definition of w, our foregoing hypothesis is inadmissible; therefore $U \mathfrak{3} \psi_o(u_1)$ and hence also $U = \psi_o(u_1)$, as stated. From α, β, γ, δ it then follows by (71) that U is a simply infinite system set in order by ψ, which was to be proved.

123. Theorem. In consequence of (121), (122) any part T of the number-series N is finite or simply infinite, according as a greatest number exists or does not exist in T.

IX.

DEFINITION OF A TRANSFORMATION OF THE NUMBER-SERIES BY INDUCTION.

124. In what follows we denote numbers by small Italics and retain throughout all symbols of the previous sections VI. to VIII., while Ω designates an arbitrary system whose elements are not necessarily contained in N.

125. Theorem. If there is given an arbitrary (similar or dissimilar) transformation θ of a system Ω in itself, and besides a determinate element ω in Ω, then to every number n corresponds one transformation ψ_n and one only of the associated number-system Z_n explained in (98), which satisfies the conditions:*

*For clearness here and in the following theorom (126) I have especially mentioned condition I., although properly it is a consequence of II. and III

I. $\psi_n(Z_n) \mathfrak{3} \Omega$

II. $\psi_n(1) = \omega$

III. $\psi_n(t') = \theta\psi_n(t)$, if $t < n$, where the symbol $\theta\psi_n$ has the meaning given in (25).

Proof by complete induction (80).

ρ. The theorem is true for $n = 1$. In this case indeed by (102) the system Z_n consists of the single number 1, and the transformation ψ, is therefore completely defined by II alone so that I is fulfilled while III drops out entirely.

σ. If the theorem is true for a number n then we show that it is also true for the following number $p = n'$, and we begin by proving that there can be only a single corresponding transformation ψ_p of the system Z_p. In fact, if a transformation ψ_p satisfies the conditions

I′. $\psi_p(Z_p) \mathfrak{3} \Omega$

II′. $\psi_p(1) = \omega$

III′. $\psi_p(m') = \theta\psi_p(m)$, when $m < p$, then there is also contained in it by (21), because $Z_n \mathfrak{3} Z_p$ (107) a transformation of Z_n which obviously satisfies the same conditions I, II, III as ψ_n, and therefore coincides throughout with ψ_n; for all numbers contained in Z_n, and hence (98) for all numbers m which are $< p$, i. e., $\leq n$, must therefore

$$\psi_p(m) = \psi_n(m) \tag{m}$$

whence there follows, as a special case,

$$\psi_p(n) = \psi_n(n); \tag{n}$$

since further by (105), (108) p is the only number of

the system Z_p not contained in Z_n, and since by III′ and (n) we must also have

$$\psi_p(p)=\theta\psi_n(n) \qquad (p)$$

there follows the correctness of our foregoing statement that there can be only one transformation ψ_p of the system Z_p satisfying the conditions I′, II′, III′, because by the conditions (m) and (p) just derived ψ_p is completely reduced to ψ_n. We have next to show conversely that this transformation ψ_p of the system Z_p completely determined by (m) and (p) actually satisfies the conditions I′, II′, III′. Obviously I′ follows from (m) and (p) with reference to I, and because $\theta(\Omega)\,3\,\Omega$. Similarly II′ follows from (m) and II, since by (99) the number 1 is contained in Z_n. The correctness of III′ follows first for those numbers m which are $< n$ from (m) and III, and for the single number $m=n$ yet remaining it results from (p) and (n). Thus it is completely established that from the validity of our theorem for the number n always follows its validity for the following number p, which was to be proved.

126. Theorem of the definition by induction. If there is given an arbitrary (similar or dissimilar) transformation θ of a system Ω in itself, and besides a determinate element ω in Ω, then there exists one and only one transformation ψ of the number-series N, which satisfies the conditions

I. $\psi(N)\,3\,\Omega$

II. $\psi(1)=\omega$

III. $\psi(n') = \theta\psi(n)$, where n represents every number.

Proof. Since, if there actually exists such a transformation ψ, there is contained in it by (21) a transformation ψ_n of the system Z_n, which satisfies the conditions I, II, III stated in (125), then because there exists one and only one such transformation ψ_n must necessarily

$$\psi(n) = \psi_n(n). \qquad (n)$$

Since thus ψ is completely determined it follows also that there can exist only one such transformation ψ (see the closing remark in (130)). That conversely the transformation ψ determined by (n) also satisfies our conditions I, II, III, follows easily from (n) with reference to the properties I, II and (p) shown in (125), which was to be proved.

127. Theorem. Under the hypotheses made in the foregoing theorem,

$$\psi(T') = \theta\psi(T),$$

where T denotes any part of the number-series N.

Proof. For if t denotes every number of the system T, then $\psi(T')$ consists of all elements $\psi(t')$, and $\theta\psi(T)$ of all elements $\theta\psi(t)$; hence our theorem follows because by III in (126) $\psi(t') = \theta\psi(t)$.

128. Theorem. If we maintain the same hypotheses and denote by θ_o the chains (44) which correspond to the transformation θ of the system Ω in itself, then is

$$\psi(N) = \theta_o(\omega).$$

Proof. We show first by complete induction (80) that

$$\psi(N) \mathfrak{3} \theta_o(\omega),$$

i. e., that every transform $\psi(n)$ is also element of $\theta_o(\omega)$. In fact,

ρ. this theorem is true for $n=1$, because by (126, II) $\psi(1)=\omega$, and because by (45) $\omega \mathfrak{3} \theta_o(\omega)$.

σ. If the theorem is true for a number n, and hence $\psi(n) \mathfrak{3} \theta_o(\omega)$, then by (55) also $\theta(\psi(n)) \mathfrak{3} \theta_o(\omega)$, i. e., by (126, III) $\psi(n') \mathfrak{3} \theta_o(\omega)$, hence the theorem is true for the following number n', which was to be proved.

In order further to show that every element ν of the chain $\theta_o(\omega)$ is contained in $\psi(N)$, therefore that

$$\theta_o(\omega) \mathfrak{3} \psi(N)$$

we likewise apply complete induction, i. e., theorem (59) transferred to Ω and the transformation θ. In fact,

ρ. the element $\omega=\psi(1)$, and hence is contained in $\psi(N)$.

σ. If ν is a common element of the chain $\theta_o(\omega)$ and the system $\psi(N)$, then $\nu=\psi(n)$, where n denotes a number, and by (126, III) we get $\theta(\nu)=\theta\psi(n)=\psi(n')$, and therefore $\theta(\nu)$ is contained in $\psi(N)$, which was to be proved.

From the theorems just established, $\psi(N) \mathfrak{3} \theta_o(\omega)$ and $\theta_o(\omega) \mathfrak{3} \psi(N)$, we get by (5) $\psi(N)=\theta_o(\omega)$, which was to be proved.

129. Theorem. Under the same hypotheses we have generally:

$$\psi(n_o) = \theta_o(\psi(n)).$$

Proof by complete induction (80). For

ρ. By (128) the theorem holds for $n=1$, since $1_o = N$ and $\psi(1) = \omega$.

σ. If the theorem is true for a number n, then

$$\theta(\psi(n_o)) = \theta(\theta_o(\psi(n)));$$

since by (127), (75)

$$\theta(\psi(n_o)) = \psi(n'_o),$$

and by (57), (126, III)

$$\theta(\theta_o(\psi(n))) = \theta_o(\theta(\psi(n))) = \theta_o(\psi(n')),$$

we get $$\psi(n'_o) = \theta_o(\psi(n')),$$

i. e., the theorem is true for the number n' following n, which was to be proved.

130. Remark. Before we pass to the most important applications of the theorem of definition by induction proved in (126), (sections X–XIV), it is worth while to call attention to a circumstance by which it is essentially distinguished from the theorem of demonstration by induction proved in (80) or rather in (59), (60), however close may seem the relation between the former and the latter. For while the theorem (59) is true quite generally for every chain A_o where A is any part of a system S transformed in itself by any transformation ϕ (IV), the case is quite different with the theorem (126), which declares only the existence of a consistent (or one-to-one) transformation ψ of the simply infinite system 1_o. If in the latter theorem (still maintaining the hypotheses regarding Ω and θ) we replace the number-series 1_o by an arbitrary

chain A_o out of such a system S, and define a transformation ψ of A_o in Ω in a manner analogous to that in (126, II, III) by assuming that

ρ. to every element a of A there is to correspond a determinate element $\psi(a)$ selected from Ω, and

σ. for every element n contained in A_o and its transform $n' = \phi(n)$, the condition $\psi(n') = \theta\psi(n)$ is to hold, then would the case very frequently occur that such a transformation ψ does not exist, since these conditions ρ, σ may prove incompatible, even though the freedom of choice contained in ρ be restricted at the outset to conform to the condition σ. An example will be sufficient to convince one of this. If the system S consisting of the different elements a and b is so transformed in itself by ϕ that $a' = b$, $b' = a$, then obviously $a_o = b_o = S$; suppose further the system Ω consisting of the different elements α, β and γ be so transformed in itself by θ that $\theta(\alpha) = \beta$, $\theta(\beta) = \gamma$, $\theta(\gamma) = \alpha$; if we now demand a transformation ψ of a_o in Ω such that $\psi(a) = \alpha$, and that besides for every element n contained in a_o always $\psi(n') = \theta\psi(n)$, we meet a contradiction; since for $n = a$, we get $\psi(b) = \theta(\alpha) = \beta$, and hence for $n = b$, we must have $\psi(a) = \theta(\beta) = \gamma$, while we had assumed $\psi(a) = \alpha$.

But if there exists a transformation ψ of A_o in Ω, which satisfies the foregoing conditions ρ, σ without contradiction, then from (60) it follows easily that it is completely determined; for if the transformation χ satisfies the same conditions, then we have, generally,

$\chi(n) = \psi(n)$, since by ρ this theorem is true for all elements $n = a$ contained in A, and since if it is true for an element n of A_0 it must by σ be true also for its transform n'.

131. In order to bring out clearly the import of our theorem (126), we will here insert a consideration which is useful for other investigations also, e. g., for the so-called group-theory.

We consider a system Ω, whose elements allow a certain combination such that from an element ν by the effect of an element ω, there always results again a determinate element of the same system Ω, which may be denoted by $\omega . \nu$ or $\omega\nu$, and in general is to be distinguished from $\nu\omega$. We can also consider this in such a way that to every determinate element ω, there corresponds a determinate transformation of the system Ω in itself (to be denoted by $\dot{\omega}$), in so far as every element ν furnishes the determinate transform $\dot{\omega}(\nu) = \omega\nu$. If to this system Ω and its element ω we apply theorem (126), designating by $\dot{\omega}$ the transformation there denoted by θ, then there corresponds to every number n a determinate element $\psi(n)$ contained in Ω, which may now be denoted by the symbol ω^n and sometimes called the nth power of ω; this notion is completely defined by the conditions imposed upon it

$$\text{II. } \omega^1 = \omega$$

$$\text{III. } \omega^{n'} = \omega\omega^n,$$

and its existence is established by the proof of theorem (126).

If the foregoing combination of the elements is further so qualified that for arbitrary elements μ, ν, ω, we always have $\omega(\nu\mu) = \omega\nu(\mu)$, then are true also the theorems

$$\omega^{n'} = \omega^n \omega, \quad \omega^m \omega^n = \omega^n \omega^m,$$

whose proofs can easily be effected by complete induction and may be left to the reader.

The foregoing general consideration may be immediately applied to the following example. If S is a system of arbitrary elements, and Ω the associated system whose elements are all the transformations ν of S in itself (36), then by (25) can these elements be continually compounded, since $\nu(S) 3 S$, and the transformation $\omega\nu$ compounded out of such transformations ν and ω is itself again an element of Ω. Then are also all elements ω^n transformations of S in itself, and we say they arise by repetition of the transformation ω. We will now call attention to a simple connection existing between this notion and the notion of the chain $\omega_o(A)$ defined in (44), where A again denotes any part of S. If for brevity we denote by A_n the transform $\omega^n(A)$ produced by the transformation ω^n, then from III and (25) it follows that $\omega(A_n) = A_{n'}$. Hence it is easily shown by complete induction (80) that all these systems A_n are parts of the chain $\omega_o(A)$; for

ρ. by (50) this statement is true for $n = 1$, and

σ. if it is true for a number n, then from (55) and from $A_{n'} = \omega(A_n)$ it follows that it is also true for the following number n', which was to be proved. Since further by (45) $A \, 3 \, \omega_o(A)$, then from (10) it results that the system K compounded out of A and all transforms A_n is part of $\omega_o(A)$. Conversely, since by (23) $\omega(K)$ is compounded out of $\omega(A) = A_1$ and all systems $\omega(A_n) = A_{n'}$, therefore by (78) out of all systems A_n, which by (9) are parts of K, then by (10) is $\omega(K) \, 3 \, K$, i. e., K is a chain (37), and since by (9) $A \, 3 \, K$, then by (47) it follows also that that $\omega_o(A) \, 3 \, K$. Therefore $\omega_o(A) = K$, i. e., the following theorem holds: If ω is a transformation of a system S in itself, and A any part of S, then is the chain of A corresponding to the transformation ω compounded out of A and all the transforms $\omega^n(A)$ resulting from repetitions of ω. We advise the reader with this conception of a chain to return to the earlier theorems (57), (58).

X.

THE CLASS OF SIMPLY INFINITE SYSTEMS.

132. Theorem. All simply infinite systems are similar to the number-series N and consequently by (33) also to one another.

Proof. Let the simply infinite system Ω be set in order (71) by the transformation θ, and let ω be the base-element of Ω thus resulting; if we again denote by θ_o the chains corresponding to the transformation θ (44), then by (71) is the following true:

α. $\theta(\Omega) \, 3 \, \Omega$.

β. $\Omega = \theta_o(\omega)$.

γ. ω is not contained in $\theta(\Omega)$.

δ. The transformation θ is similar.

If then ψ denotes the transformation of the number-series N defined in (126), then from β and (128) we get first

$$\psi(N)=\Omega,$$

and hence we have only yet to show that ψ is a similar transformation, i. e., (26) that to different numbers m, n correspond different transforms $\psi(m)$, $\psi(n)$. On account of the symmetry we may by (90) assume that $m>n$, hence $m \ni n'_o$, and the theorem to prove comes to this that $\psi(n)$ is not contained in $\psi(n'_o)$, and hence by (127) is not contained in $\theta\psi(n_o)$. This we establish for every number n by complete induction (80). In fact,

ρ. this theorem is true by γ for $n=1$, since $\psi(1)=\omega$ and $\psi(1_o)=\psi(N)=\Omega$.

σ. If the theorem is true for a number n, then is it also true for the following number n'; for if $\psi(n')$, i. e., $\theta\psi(n)$, were contained in $\theta\psi(n'_o)$, then by δ and (27), $\psi(n)$ would also be contained in $\psi(n'_o)$ while our hypothesis states just the opposite; which was to be proved.

133. Theorem. Every system which is similar to a simply infinite system and therefore by (132), (33) to the number-series N is simply infinite.

Proof. If Ω is a system similar to the number-series N, then by (32) there exists a similar transformation ψ of N such that

$$\text{I.}\quad \psi(N) = \Omega;$$

then we put

$$\text{II.}\quad \psi(1) = \omega.$$

If we denote, as in (26), by $\bar{\psi}$ the inverse, likewise similar transformation of Ω, then to every element ν of Ω there corresponds a determinate number $\bar{\psi}(\nu) = n$, viz., that number whose transform $\psi(n) = \nu$. Since to this number n there corresponds a determinate following number $\phi(n) = n'$, and to this again a determinate element $\psi(n')$ in Ω there belongs to every element ν of the system Ω a determinate element $\psi(n')$ of that system which as transform of ν we shall designate by $\theta(\nu)$. Thus a transformation θ of Ω in itself is completely determined,* and in order to prove our theorem we will show that by θ Ω is set in order (71) as a simply infinite system, i. e., that the conditions α, β, γ, δ stated in the proof of (132) are all fulfilled. First α is immediately obvious from the definition of θ. Since further to every number n corresponds an element $\nu = \phi(n)$, for which $\theta(\nu) = \psi(n')$, we have generally,

$$\text{III.}\quad \psi(n') = \theta\psi(n),$$

and thence in connection with I, II, α it results that the transformations θ, ψ fulfill all the conditions of theorem (126); therefore β follows from (128) and I. Further by (127) and I

$$\psi(N') = \theta\psi(N) = \theta(\Omega),$$

and thence in combination with II and the similarity

* Evidently θ is the transformation $\psi\phi\bar{\psi}$ compounded by (25) out of $\bar{\psi}$, ϕ, ψ.

of the transformation ψ follows γ, because otherwise $\psi(1)$ must be contained in $\psi(N')$, hence by (27) the number 1 in N', which by (71, γ) is not the case. If finally μ, ν denote elements of Ω and m, n the corresponding numbers whose transforms are $\psi(m)=\mu$, $\psi(n)=\nu$, then from the hypothesis $\theta(\mu)=\theta(\nu)$ it follows by the foregoing that $\psi(m')=\psi(n')$, thence on account of the similarity of ψ, ϕ that $m'=n'$, $m=n$, therefore also $\mu=\nu$; hence also δ is true, which was to be proved.

134. Remark. By the two preceding theorems (132), (133) all simply infinite systems form a class in the sense of (34). At the same time, with reference to (71), (73) it is clear that every theorem regarding numbers, i. e., regarding the elements n of the simply infinite system N set in order by the transformation ϕ' and indeed every theorem in which we leave entirely out of consideration the special character of the elements n and discuss only such notions as arise from the arrangement ϕ, possesses perfectly general validity for every other simply infinite system Ω set in order by a transformation θ and its elements ν, and that the passage from N to Ω (e. g., also the translation of an arithmetic theorem from one language into another) is effected by the transformation ψ considered in (132), (133), which changes every element n of N into an element ν of Ω, i. e., into $\psi(n)$. This element ν can be called the nth element of Ω and accordingly the number n is itself the nth number of the number-

series N. The same significance which the transformation ϕ possesses for the laws in the domain N, in so far as every element n is followed by a determinate element $\phi(n)=n'$, is found, after the change effected by ψ, to belong to the transformation θ for the same laws in the domain Ω, in so far as the element $\nu=\psi(n)$ arising from the change of n is followed by the element $\theta(\nu)=\psi(n')$ arising from the change of n'; we are therefore justified in saying that by ψ ϕ is changed into θ, which is symbolically expressed by $\theta=\psi\phi\bar{\psi}$, $\phi=\bar{\psi}\theta\psi$. By these remarks, as I believe, the definition of the notion of numbers given in (73) is fully justified. We now proceed to further applications of theorem (126).

XI.

ADDITION OF NUMBERS.

135. Definition. It is natural to apply the definition set forth in theorem (126) of a transformation ψ of the number-series N, or of the *function* $\psi(n)$ determined by it to the case, where the system there denoted by Ω in which the transform $\psi(N)$ is to be contained, is the number-series N itself, because for this system Ω a transformation θ of Ω in itself already exists, viz., that transformation ϕ by which N is set in order as a simply infinite system (71), (73). Then is also $\Omega=N$, $\theta(n)=\phi(n)=n'$, hence

I. $\psi(N) \mathrel{3} N$,

and it remains in order to determine ψ completely

only to select the element ω from Ω, i. e., from N, at pleasure. If we take $\omega = 1$, then evidently ψ becomes the identical transformation (21) of N, because the conditions

$$\psi(1) = 1, \quad \psi(n') = (\psi(n))'$$

are generally satisfied by $\psi(n) = n$. If then we are to produce another transformation ψ of N, then for ω we must select a number m' different from 1, by (78) contained in N, where m itself denotes any number; since the transformation ψ is obviously dependent upon the choice of this number m, we denote the corresponding transform $\psi(n)$ of an arbitrary number n by the symbol $m + n$, and call this number the *sum* which arises from the number m by the *addition* of the number n, or in short the sum of the numbers m, n. Therefore by (126) this sum is completely determined by the conditions*

$$\text{II.} \quad m + 1 = m',$$

$$\text{III.} \quad m + n' = (m + n)'.$$

136. Theorem. $m' + n = m + n'$.

Proof by complete induction (80). For

ρ. the theorem is true for $n = 1$, since by (135, II)

$$m' + 1 = (m')' = (m + 1)',$$

and by (135, III) $(m + 1)' = m + 1'$.

* The above definition of addition based immediately upon theorem (126) seems to me to be the simplest. By the aid of the notion developed in (131) we can, however, define the sum $m+n$ by $\phi^n(m)$ or also by $\phi^m(n)$, where ϕ has again the foregoing meaning. In order to show the complete agreement of these definitions with the foregoing, we need by (126) only to show that if $\phi^n(m)$ or $\phi^m(n)$ is denoted by $\psi(n)$, the conditions $\psi(1) = m'$, $\psi(n') = \phi\psi(n)$ are fulfilled which is easily done with the aid of complete induction (80) by the help of (131).

σ. If the theorem is true for a number n, and we put the following number $n' = p$, then is $m' + n = m + p$, hence also $(m' + n)' = (m + p)'$, whence by (135, III) $m' + p = m + p'$; therefore the theorem is true also for the following number p, which was to be proved.

137. Theorem. $m' + n = (m + n)'$.

The proof follows from (136) and (135, III).

138. Theorem. $1 + n = n'$.

Proof by complete induction (80). For

ρ. by (135, II) the theorem is true for $n = 1$.

σ. If the theorem is true for a number n and we put $n' = p$, then $1 + n = p$, therefore also $(1 + n)' = p'$, whence by (135, III) $1 + p = p'$, i. e., the theorem is true also for the following number p, which was to be proved.

139. Theorem. $1 + n = n + 1$.

The proof follows from (138) and (135, II).

140. Theorem. $m + n = n + m$.

Proof by complete induction (80). For

ρ. by (139) the theorem is true for $n = 1$.

σ. If the theorem is true for a number n, then there follows also $(m + n)' = (n + m)'$, i. e., by (135, III) $m + n' = n + m'$, hence by (136) $m + n' = n' + m$; therefore the theorem is also true for the following number n', which was to be proved.

141. Theorem. $(l + m) + n = l + (m + n)$.

Proof by complete induction (80). For

ρ. the theorem is true for $n=1$, because by (135, II, III, II) $(l+m)+1=(l+m)'=l+m'=l+(m+1)$.

σ. If the theorem is true for a number n, then there follows also $((l+m)+n)'=(l+(m+n))'$, i. e., by (135, III)

$$(l+m)+n'=l+(m+n)'=l+(m+n'),$$

therefore the theorem is also true for the following number n', which was to be proved.

142. Theorem. $m+n>m$.

Proof by complete induction (80). For

ρ. by (135, II) and (91) the theorem is true for $n=1$.

σ. If the theorem is true for a number n, then by (95) it is also true for the following number n', because by (135, III) and (91)

$$m+n'=(m+n)'>m+n,$$

which was to be proved.

143. Theorem. The conditions $m>a$ and $m+n>a+n$ are equivalent.

Proof by complete induction (80). For

ρ. by (135, II) and (94) the theorem is true for $n=1$.

σ. If the theorem is true for a number n, then is it also true for the following number n', since by (94) the condition $m+n>a+n$ is equivalent to $(m+n)'>(a+n)'$, hence by (135, III) also equivalent to

$$m+n'>a+n',$$

which was to be proved.

144. Theorem. If $m > a$ and $n > b$, then is also

$$m + n > a + b.$$

Proof. For from our hypotheses we have by (143) $m + n > a + n$ and $n + a > b + a$ or, what by (140) is the same, $a + n > a + b$, whence the theorem follows by (95).

145. Theorem. If $m + n = a + n$, then $m = a$.

Proof. For if m does not $= a$, hence by (90) either $m > a$ or $m < a$, then by (143) respectively $m + n > a + n$ or $m + n < a + n$, therefore by (90) we surely cannot have $m + n = a + n$, which was to be proved.

146. Theorem. If $l > n$, then there exists one and by (157) only one number m which satisfies the condition $m + n = l$.

Proof by complete induction (80). For

ρ. the theorem is true for $n = 1$. In fact, if $l > 1$, i. e., (89) if l is contained in N', and hence is the transform m' of a number m, then by (135, II) it follows that $l = m + 1$, which was to be proved.

σ. If the theorem is true for a number n, then we show that it is also true for the following number n'. In fact, if $l > n'$, then by (91), (95) also $l > n$, and hence there exists a number k which satisfies the condition $l = k + n$; since by (138) this is different from 1 (otherwise l would be $= n'$) then by (78) is it the transform m' of a number m, consequently $l = m' + n$, therefore also by (136) $l = m + n'$, which was to be proved.

XII.

MULTIPLICATION OF NUMBERS.

147. Definition. After having found in XI an infinite system of new transformations of the number-series N in itself, we can by (126) use each of these in order to produce new transformations ψ of N. When we take $\Omega = N$, and $\theta(n) = m + n = n + m$, where m is a determinate number, we certainly again have

$$\text{I.} \quad \psi(N) \mathfrak{3} N,$$

and it remains, to determine ψ completely only to select the element ω from N at pleasure. The simplest case occurs when we bring this choice into a certain agreement with the choice of θ, by putting $\omega = m$. Since the thus perfectly determinate ψ depends upon this number m, we designate the corresponding transform $\psi(n)$ of any number n by the symbol $m \times n$ or $m.n$ or mn, and call this number the *product* arising from the number m by *multiplication* by the number n, or in short the product of the numbers m, n. This therefore by (126) is completely determined by the conditions

$$\text{II.} \quad m.1 = m$$

$$\text{III.} \quad mn' = mn + m,$$

148. Theorem. $m'n = mn + n$.

Proof by complete induction (80). For

ρ. by (147, II) and (135, II) the theorem is true for $n = 1$.

σ. If the theorem is true for a number n, we have

$$m'n + m' = (mn + n) + m'$$

and consequently by (147, III), (141), (140), (136), (141), (147, III)

$$m'n' = mn + (n + m') = mn + (m' + n) = mn + (m + n')$$
$$= (mn + m) + n' = mn' + n';$$

therefore the theorem is true for the following number n', which was to be proved.

149. Theorem. $1 . n = n$.

Proof by complete induction (80). For

ρ. by (147, II) the theorem is true for $n = 1$.

σ. If the theorem is true for a number n, then we have $1 . n + 1 = n + 1$, i. e., by (147, III), (135, II) $1 . n' = n'$, therefore the theorem also holds for the following number n', which was to be proved.

150. Theorem. $mn = nm$.

Proof by complete induction (80). For

ρ by (147, II), (149) the theorem is true for $n = 1$.

σ. If the theorem is true for a number n, then we have

$$mn + m = nm + m,$$

i. e., by (147, III), (148) $mn' = n'm$, therefore the theorem is also true for the following number n', which was to be proved.

151. Theorem. $l(m + n) = lm + ln$.

Proof by complete induction (80). For

ρ. by (135, II), (147, III), (147, II) the theorem is true for $n = 1$.

σ. If the theorem is true for a number n, we have

$$l(m+n)+l=(lm+ln)+l;$$

but by (147, III), (135, III) we have

$$l(m+n)+l=l(m+n)'=l(m+n'),$$

and by (141), (147, III)

$$(lm+ln)+l=lm+(ln+l)=lm+ln',$$

consequently $l(m+n')=lm+ln'$, i. e., the theorem is true also for the following number n', which was to be proved.

152. Theorem. $(m+n)l=ml+nl$.

The proof follows from (151), (150).

153. Theorem. $(lm)n=l(mn)$.

Proof by complete induction (80). For

ρ. by (147, II) the theorem is true for $n=1$.

σ. If the theorem is true for a number n, then we have

$$(lm)n+lm=l(mn)+lm,$$

i. e., by (147, III), (151), (147, III)

$$(lm)n'=l(mn+m)=l(mn'),$$

hence the theorem is also true for the following number n', which was to be proved.

154. Remark. If in (147) we had assumed no relation between ω and θ, but had put $\omega=k$, $\theta(n)=m+n$, then by (126) we should have had a less simple transformation ψ of the number-series N; for the number 1 would $\psi(1)=k$ and for every other number (therefore contained in the form n') would $\psi(n')=mn+k$; since thus would be fulfilled, as one could

easily convince himself by the aid of the foregoing theorems, the condition $\psi(n') = \theta\psi(n)$, i. e., $\psi(n') = m + \psi(n)$ for all numbers n.

XIII.

INVOLUTION OF NUMBERS.

155. Definition. If in theorem (126) we again put $\Omega = N$, and further $\omega = a$, $\theta(n) = an = na$, we get a transformation ψ of N which still satisfies the condition

$$\text{I. } \psi(N) \,3\, N;$$

the corresponding transform $\psi(n)$ of any number n we denote by the symbol a^n, and call this number a *power of the base* a, while n is called the *exponent* of this power of a. Hence this notion is completely determined by the conditions

$$\text{II. } a^1 = a$$

$$\text{III. } a^{n'} = a \,.\, a^n = a^n \,.\, a.$$

156. Theorem. $a^{m+n} = a^m \,.\, a^n$.

Proof by complete induction (80). For

ρ. by (135, II), (155, III), (155, II) the theorem is true for $n = 1$.

σ. If the theorem is true for a number n, we have

$$a^{m+n} \,.\, a = (a^m \,.\, a^n)a;$$

but by (155, III), (135, III) $a^{m+n} \,.\, a = a^{(m+n)'} = a^{m+n'}$, and by (153), (155, III) $(a^m \,.\, a^n)\,a = a^m(a^n \,.\, a) = a^m \,.\, a^{n'}$; hence $a^{m+n'} = a^m \,.\, a^{n'}$, i. e., the theorem is also true for the following number n', which was to be proved.

157. Theorem. $(a^m)^n = a^{mn}$.

Proof by complete induction (80). For

ρ. by (155, II), (147, II) the theorem is true for $n=1$.

σ. If the theorem is true for a number n, we have

$$(a^m)^n \,.\, a^m = a^{mn} \,.\, a^m$$

but by (155, III) $(a^m)^n \,.\, a^m = (a^m)^{n'}$, and by (156), (147, III) $a^{mn} \,.\, a^m = a^{mn+m} = a^{mn'}$; hence $(a^m)^{n'} = a^{mn'}$, i. e., the theorem is also true for the following number n', which was to be proved.

158. Theorem. $(a\,b)^n = a^n \,.\, b^n$.

Proof by complete induction (80). For

ρ. by (155, II) the theorem is true for $n=1$.

σ. If the theorem is true for a number n, then by (150), (153), (155, III) we have also $(a\,b)^n . a = a(a^n . b^n) = (a . a^n)\, b^n = a^{n'} . b^n$, and thus $((a\,b)^n . a)\, b = (a^{n'} . b^n) b$; but by (153), (155, III) $((a\,b)^n . a)\, b = (a\,b)^n . (a\,b) = (a\,b)^{n'}$, and likewise

$$(a^{n'} . b^n)\, b = a^{n'} . (b^n . b) = a^{n'} . b^{n'};$$

therefore $(a\,b)^{n'} = a^{n'} . b^{n'}$, i. e., the theorem is also true for the following number n', which was to be proved.

XIV.

NUMBER OF THE ELEMENTS OF A FINITE SYSTEM.

159. Theorem. If Σ is an infinite system, then is every one of the number-systems Z_n defined in (98) similarly transformable in Σ (i. e., similar to a part of Σ), and conversely.

Proof. If Σ is infinite, then by (72) there certainly exists a part T of Σ, which is simply infinite, therefore by (132) similar to the number-series N, and consequently by (35) every system Z_n as part of N is similar to a part of T, therefore also to a part of Σ, which was to be proved.

The proof of the converse—however obvious it may appear—is more complicated. If every system Z_n is similarly transformable in Σ, then to every number n corresponds such a similar transformation α_n of Z_n that $\alpha_n(Z_n) 3 \Sigma$. From the existence of such a series of transformations α_n, regarded as given, but respecting which nothing further is assumed, we derive first by the aid of theorem (126) the existence of a new series of such transformations ψ_n possessing the special property that whenever $m \leqq n$, hence by (100) $Z_m 3 Z_n$, the transformation ψ_m of the part Z_m is contained in the transformation ψ_n of Z_n (21), i. e., the transformations ψ_m and ψ_n completely coincide with each other for all numbers contained in Z_m, hence always

$$\psi_m(m) = \psi_n(m).$$

In order to apply the theorem stated to gain this end we understand by Ω that system whose elements are all possible similar transformations of all systems Z_n in Σ, and by aid of the given elements α_n, likewise contained in Ω, we define in the following manner a transformation θ of Ω in itself. If β is any element of Ω, thus, e. g., a similar transformation of the determinate system Z_n in Σ, then the system $\alpha_{n'}(Z_{n'})$ cannot be part of $\beta(Z_n)$, for otherwise $Z_{n'}$ would be similar by (35) to a part of Z_n, hence by (107) to a proper part of itself, and consequently infinite, which would contradict theorem (119); therefore there certainly exists in $Z_{n'}$ one number or several numbers p such that $\alpha_{n'}(p)$ is not contained in $\beta(Z_n)$; from these numbers p we select—simply to lay down something determinate—always the least k (96) and, since $Z_{n'}$ by (108) is compounded out of Z_n and n', define a transformation γ of $Z_{n'}$ such that for all numbers m contained in Z_n the transform $\gamma(m)=\beta(m)$ and besides $\gamma(n')=\alpha_{n'}(k)$; this obviously similar transformation γ of $Z_{n'}$ in Σ we consider then as a transform $\theta(\beta)$ of the transformation β, and thus a transformation θ of the system Ω in itself is completely defined. After the things named Ω and θ in (126) are determined we select finally for the element of Ω denoted by ω the given transformation α_1; thus by (126) there is determined a transformation ψ of the number-series N in Ω, which, if we denote the transform belonging to an arbitrary number n, not by $\psi(n)$ but by ψ_n, satisfies the conditions

$$\text{II. } \psi_1=\alpha_1$$

$$\text{III. } \psi_{n'}=\theta(\psi_n)$$

By complete induction (80) it results first that ψ_n is a similar transformation of Z_n in Σ; for

ρ. by II this is true for $n=1$.

σ. if this statement is true for a number n, it follows from III and from the character of the above described transition θ from β to γ, that the statement is also true for the following number n', which was to be proved. Afterward we show likewise by complete induction (80) that if m is any number the above stated property

$$\psi_n(m) = \psi_m(m)$$

actually belongs to all numbers n, which are $\geqq m$, and therefore by (93), (74) belong to the chain m_o; in fact,

ρ. this is immediately evident for $n = m$, and

σ. if this property belongs to a number n it follows again from III and the nature of θ, that it also belongs to the number n', which was to be proved. After this special property of our new series of transformations ψ_n has been established, we can easily prove our theorem. We define a transformation χ of the number-series N, in which to every number n we let the transform $\chi(n) = \psi_n(n)$ correspond; obviously by (21) all transformations ψ_n are contained in this one transformation χ. Since ψ_n was a transformation of Z_n in Σ, it follows first that the number series N is likewise transformed by χ in Σ, hence $\chi(N) 3 \Sigma$. If further m, n are different numbers we may by reason of symmetry according to (90) suppose $m < n$; then by the foregoing $\chi(m) = \psi_m(m) = \psi_n(m)$, and $\chi(n) = \psi_n(n)$; but since ψ_n was a similar transformation of Z_n in Σ, and m, n are different elements of Z_n, then is $\psi_n(m)$ different from $\psi_n(n)$, hence also $\chi(m)$ different from $\chi(n)$, i. e., χ is a similar transformation of N. Since further N is an infinite system (71), the same thing is true by (67) of the system $\chi(N)$ similar to it and by (68), because $\chi(N)$ is part of Σ, also of Σ, which was to be proved.

160. Theorem. A system Σ is finite or infinite, according as there does or does not exist a system Z_n similar to it.

Proof. If Σ is finite, then by (159) there exist systems Z_n which are not similarly transformable in Σ; since by (102) the system Z_1 consists of the single number 1, and hence is similarly transformable in every system, then must the least number k (96) to which a system Z_k not similarly transformable in Σ corresponds be different from 1 and hence by (78) $=n'$, and since $n < n'$ (91) there exists a similar transformation ψ of Z_n in Σ; if then $\psi(Z_n)$ were only a proper part of Σ, i. e., if there existed an element α in Σ not contained in $\psi(Z_n)$, then since $Z_{n'} = \mathfrak{M}(Z_n, n')$ (108) we could extend this transformation ψ to a similar transformation ψ of $Z_{n'}$ in Σ by putting $\psi(n') = \alpha$, while by our hypothesis $Z_{n'}$ is not similarly transformable in Σ. Hence $\psi(Z_n) = \Sigma$, i. e., Z_n and Σ are similar systems. Conversely, if a system Σ is similar to a system Z_n, then by (119), (67) Σ is finite, which was to be proved.

161. Definition. If Σ is a finite system, then by (160) there exists one and by (120), (33) only one single number n to which a system Z_n similar to the system Σ corresponds; this number n is called the *number* [*Anzahl*] of the elements contained in Σ (or

also the *degree* of the system Σ) and we say Σ consists of or is a system of n elements, or the number n shows *how many* elements are contained in Σ.* If numbers are used to express accurately this determinate property of finite systems they are called *cardinal numbers*. As soon as a determinate similar transformation ψ of the system Z_n is chosen by reason of which $\psi(Z_n)=Z$, then to every number m contained in Z_n (i. e., every number m which is $\leqq n$) there corresponds a determinate element $\psi(m)$ of the system Σ, and conversely by (26) to every element of Σ by the inverse transformation $\bar{\psi}$ there corresponds a determinate number m in Z_n. Very often we denote all elements of Σ by a single letter, e. g., α, to which we append the distinguishing numbers m as indices so that $\psi(m)$ is denoted by α_m. We say also that these elements are *counted and set in order* by ψ in determinate manner, and call α_m the mth element of Σ; if $m < n$ then $\alpha_{m'}$ is called the element *following* α_m, and α_n is called the *last* element. In this counting of the elements therefore the numbers m appear again as ordinal numbers (73).

162. Theorem. All systems similar to a finite system possess the same number of elements.

The proof follows immediately from (33), (161).

163. Theorem. The number of numbers contained in Z_n, i. e., of those numbers which are $\leqq n$, is n.

*For clearness and simplicity in what follows we restrict the notion of the number throughout to finite systems; if then we speak of a number of certain things, it is always understood that the system whose elements these things are is a finite system.

Proof. For by (32) Z_n is similar to itself.

164. Theorem. If a system consists of a single element, then is the number of its elements $=1$, and conversely.

The proof follows immediately from (2), (26), (32), (102), (161).

165. Theorem. If T is proper part of a finite system Σ, then is the number of the elements of T less than that of the elements of Σ.

Proof. By (68) T is a finite system, therefore similar to a system Z_m, where m denotes the number of the elements of T; if further n is the number of elements of Σ, therefore Σ similar to Z_n, then by (35) T is similar to a proper part E of Z_n and by (33) also Z_m and E are similar to each other; if then we were to have $n \leqq m$, hence $Z_n \, 3 \, Z_m$, by (7) E would also be proper part of Z_m, and consequently Z_m an infinite system, which contradicts theorem (119); hence by (90), $m < n$, which was to be proved.

166. Theorem. If $\Gamma = \mathfrak{M}(B, \gamma)$, where B denotes a system of n elements, and γ an element of Γ not contained in B, then Γ consists of n' elements.

Proof. For if $B = \psi(Z_n)$, where ψ denotes a similar transformation of Z_n, then by (105), (108) it may be extended to a similar transformation ψ of $Z_{n'}$, by putting $\psi(n') = \gamma$, and we get $\psi(Z_{n'}) = \Gamma$, which was to be proved.

167. Theorem. If γ is an element of a system Γ

consisting of n' elements, then is n the number of all other elements of Γ.

Proof. For if B denotes the aggregate of all elements in Γ different from γ, then is $\Gamma = \mathfrak{M}(B, \gamma)$; if then b is the number of elements of the finite system B, by the foregoing theorem b' is the number of elements of Γ, therefore $= n'$, whence by (26) we get $b = n$, which was to be proved.

168. Theorem. If A consists of m elements, and B of n elements, and A and B have no common element, then $\mathfrak{M}(A, B)$ consists of $m + n$ elements.

Proof by complete induction (80). For

ρ. by (166), (164), (135, II) the theorem is true for $n = 1$.

σ. If the theorem is true for a number n, then is it also true for the following number n'. In fact, if Γ is a system of n' elements, then by (167) we can put $\Gamma = \mathfrak{M}(B, \gamma)$ where γ denotes an element and B the system of the n other elements of Γ. If then A is a system of m elements each of which is not contained in Γ, therefore also not contained in B, and we put $\mathfrak{M}(A, B) = \Sigma$, by our hypothesis $m + n$ is the number of elements of Σ, and since γ is not contained in Σ, then by (166) the number of elements contained in $\mathfrak{M}(\Sigma, \gamma) = (m + n)'$, therefore by (135, III) $= m + n'$; but since by (15) obviously $\mathfrak{M}(\Sigma, \gamma) = \mathfrak{M}(A, B, \gamma) = \mathfrak{M}(A, \Gamma)$, then is $m + n'$ the number of the elements of $\mathfrak{M}(A, \Gamma)$, which was to be proved.

169. Theorem. If A, B are finite systems of m, n

elements respectively, then is $\mathfrak{M}(A, B)$ a finite system and the number of its elements is $\leqq m+n$.

Proof. If $B \mathrel{3} A$, then $\mathfrak{M}(A, B)=A$, and the number m of the elements of this system is by (142) $< m+n$, as was stated. But if B is not part of A, and T is the system of all those elements of B that are not contained in A, then by (165) is their number $p \leqq n$, and since obviously

$$\mathfrak{M}(A, B)=\mathfrak{M}(A, T),$$

then by (143) is the number $m+p$ of the elements of this system $\leqq m+n$, which was to be proved.

170. Theorem. Every system compounded out of a number n of finite systems is finite.

Proof by complete induction (80). For

ρ. by (8) the theorem is self-evident for $n=1$.

σ. If the theorem is true for a number n, and if Σ is compounded out of n' finite systems, then let A be one of these systems and B the system compounded out of all the rest; since their number by (167) $=n$, then by our hypothesis B is a finite system. Since obviously $\Sigma=\mathfrak{M}(A, B)$, it follows from this and from (169) that Σ is also a finite system, which was to be proved.

171. Theorem. If ψ is a dissimilar transformation of a finite system Σ of n elements, then is the number of elements of the transform $\psi(\Sigma)$ less than n.

Proof. If we select from all those elements of Σ that possess one and the same transform, always one and only one at pleasure, then is the system T of all these selected elements obviously a proper part of Σ, because ψ is a dissimilar transformation of Σ (26). At the same time it is clear that the transformation by (21) contained in ψ of this part T is a similar transformation, and that $\psi(T) = \psi(\Sigma)$; hence the system $\psi(\Sigma)$ is similar to the proper part T of Σ, and consequently our theorem follows by (162), (165).

172. Final remark. Although it has just been shown that the number m of the elements of $\psi(\Sigma)$ is less than the number n of the elements of Σ, yet in many cases we like to say that the number of elements of $\psi(\Sigma) = n$. The word number is then, of course, used in a different sense from that used hitherto (161); for if α is an element of Σ and a the number of all those elements of Σ, that possess one and the same transform $\psi(\alpha)$ then is the latter as element of $\psi(\Sigma)$ frequently regarded still as representative of a elements, which at least from their derivation may be considered as different from one another, and accordingly counted as a-fold element of $\psi(\Sigma)$. In this way we reach the notion, very useful in many cases, of systems in which every element is endowed with a certain frequency-number which indicates how often it is to be reckoned as element of the system. In the foregoing case, e. g., we would say that n is the number of the elements of $\psi(\Sigma)$ counted in this sense, while the number m of the actually different elements of this system coincides with the number of the elements of T. Similar deviations from the orig-

inal meaning of a technical term which are simply extensions of the original notion, occur very frequently in mathematics; but it does not lie in the line of this memoir to go further into their discussion.

A CATALOG OF SELECTED

DOVER BOOKS

IN ALL FIELDS OF INTEREST

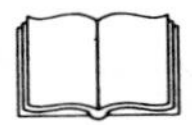

A CATALOG OF SELECTED DOVER BOOKS IN ALL FIELDS OF INTEREST